24-95
R

CW01654059

EUCALYPTUS

EUCALYPTUS

An illustrated guide to identification

Salmon gum, ***Eucalyptus salmonophloia***

IAN BROOKER AND DAVID KLEINIG

REED

First published in 1996 by
REED BOOKS
a part of Reed Books Australia
22 Salmon Street, Port Melbourne, Victoria 3207
a division of Reed International Books Australia Pty Ltd

Copyright © M.I.H. Brooker and D.A. Kleinig 1996

All rights reserved. No part of this publication may be reproduced, stored in a retrieval system or transmitted in any form or by any means, electronic, mechanical, photocopying, recording or otherwise, without the prior written permission of the publishers.

National Library of Australia
Cataloguing-in-Publication Data

Brooker, Ian.
Eucalyptus.

Bibliography.
Includes index.
ISBN 0 7301 0494 X

1. Eucalyptus - Australia - Identification.
I. Kleinig, David. II. Title.

583.420994

Layout and setting by Trevor Hood
ANACONDA GRAPHIC DESIGN

Produced in Hong Kong by Mandarin

Contents

Introduction

The first eucalypts known to science were collected in 1770 by Joseph Banks and Daniel Solander from eastern Australia. However, the name *Eucalyptus* was not coined until 1788, when the French botanist L'Heritier, working in London, erected *Eucalyptus obliqua*. This was based on a specimen of a tree collected in 1777 by David Nelson on Bruny Island (southern Tasmania) on James Cook's third voyage.

Since then hundreds of eucalypt species have been discovered and named from all over the continent and even from New Guinea and other islands north of Australia. Eucalypts are now well known in many other parts of the world, as they have been propagated for timber, fibre, fuel and ornamental purposes in many tropical and subtropical countries.

Because of their diversity in myriad features such as form, height, leaf and flower colour and adaptability, they continue to be in constant demand. It is essential that we have a system of naming and also keys to the recognition of species. Innumerable books are devoted to eucalypts, whether simply to show their beauty or to demonstrate their utility. Many of these use traditional keys to the identification of species. All employ keys based on the printed word, usually supplemented by illustrations. The disadvantage of these systems is that the user is required to proceed along a fixed path, which requires a prescribed set of characters of the species to be known. If the character is unknown, the user is stranded.

Leonard Cronin in his *Key Guide to Australian Wildflowers* devised a different form of key. In this, wildflowers are categorised according to shape, colour, etc., which are explicitly illustrated. The user recognises aspects of the flower at hand and is then directed to pages in the book dealing with all species that fall into the various categories.

We have adapted this system to the eucalypts, which are traditionally identified by quite a different set of characters—namely, tree form, bark type, flower cluster type, flower numbers per individual clusters and leaf colour, as well as the easy-to-recognise feature of the flower colour. Flowers are not always available for identification purposes, because they occur at different times of the year in different species. In this book we do not use them in the key, but they are mentioned in the digests. It is hoped that the other characters will provide sufficient information for identification to proceed.

We have added one highly informative feature. These are the maps for the natural distribution of the 200 species included in the key. They are useful if the specimen in question has been sampled from a natural population. If from a cultivated specimen, the maps cannot be used in identification and the user must rely on the botanical features alone. Nevertheless, they remain a valuable source of information about the species, particularly if you want to estimate the chances of success in planting a desired species in a region far from its natural habitat.

We have made a selection of species that consists of approximately a quarter of the whole genus. They come from every state in Australia and include the most important commercial species plus those that are most often planted for ornamental purposes. For more information about them, the reader should consult the books listed on page 222.

How to use this guide

This book is designed to help the user to identify the most commonly encountered eucalypts. There are several obvious characters that are conventionally used for eucalypts. Some understanding of their value and reliability is important. Parts of the plant that are continually exposed to the elements are, over a season, likely to be modified in appearance. A prime example is bark, which we regard as very helpful but must be used with caution. Other factors that must be taken into account are the age of the tree and the season when features are assessed. When we describe bark we mean that on a mature specimen. When we refer to leaf colour we mean the mature crown colour, not the colour of the fresh new growth even if it is on a mature tree.

Assessment of a eucalypt should be made in approximately the following order:

- the size and habit (tree or mallee)
- the colour of the crown
- the nature of the crown (i.e. is it composed of juvenile or adult leaves?)
- the bark (rough or smooth)
- the features of the individual leaves (size, shape, colour)
- position of the bud clusters (i.e. are they inside the crown, in the leaf axils or at the leafless ends of the branchlets?)
- are the bud clusters single, double or compound?
- the number of buds in a single cluster
- the bud shape
- the flower colour
- the fruit shape
- the seed

Perhaps the most difficult character to come to terms with is the bark. Most keys require a recognition of smooth or rough, then the placement of the rough forms into so-called distinct categories. The easiest to recognise are ironbark, tessellated bark and stringybark, in that order. We simply use 'rough' v. 'smooth', although at the end of one section we resort to an estimation of the approximate amount of rough bark.

The bud clusters may need explanation. In most species the bud clusters occur in the leaf axils as a simple stalk surmounted by the buds in various numbers. Other species have the clusters in leafless sprays at the ends of the branchlets (bloodwoods and boxes). A few have small leafless shoots bearing clusters in the leaf axils (ghost gums and a few others). Sugar gum is notable for having the clusters on leafless branchlets inside the current leafy crown.

A note of caution must also be given about bud numbers. Eucalypts have their buds singly, in 3s, 7s, or in >7s. There can be >50. The numbers used in the key refer always to an intact bud cluster. During a growth season, buds may abort, obscuring the real number. Usually, however, scars are left at the expanded top of the common bud stalk, and the original number can be inferred.

The key begins with 'bud numbers', as these are usually always present. Then the four categories each divide into two, depending on whether the plant is a tree or mallee. Bark type and fruit size follow. When numbers of species become manageable, final consultation is to be made by seeing all the confirmatory information in the digests, including the natural distribution.

1. NUMBER OF BUDS PER INFLORESCENCE

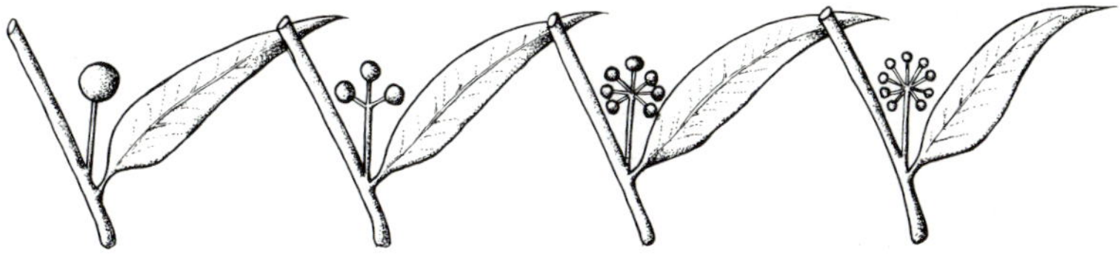

1 3 7 >7

2. FORM OF PLANT

forest tree

woodland tree

mallet

mallee

shrub

3. POSITION OF BUDS

in terminal clusters

in axillary clusters

3. POSITION OF BUDS

in paired axillary clusters

in clusters on leaf-less branchlets

4. TYPES OF BARK

rough bark

stringy

ironbark

red bloodwood

red bloodwood

yellow bloodwood

fibrous

box

box

smooth bark

4. TYPES OF BARK

smooth bark

minniritchi

other smooth barks, with

scribbles

flakes

ribbons

5. FRUIT WIDTH

large, wider than 1.5 cm

medium, 0.5–1.5 cm

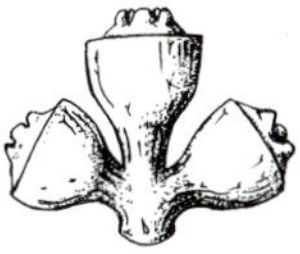

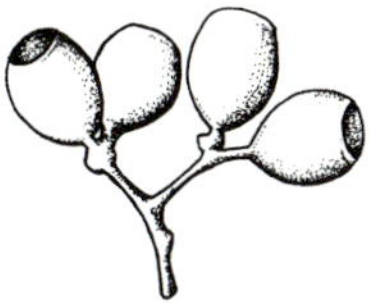

small, narrower than 0.5 cm

6. CROWN COLOUR

grey, blue-grey, grey-green, waxy-white

6. CROWN COLOUR

green or olive green

7. LEAF DISCOLOUREDNESS

discolorous

i.e. different colour top and underside

concolorous

i.e. same colour both sides

8. ROUGH BARK ON LOWER HALF OF TRUNK ONLY

9. FRUIT RIM

with steeply ascending disc

without ascending disc

10. OPERCULUM LENGTH

long

short

Characters not in the key as not always available, but referred to in every digest

11. SEEDLING AND JUVENILE LEAF STALK

present

absent

12. FLOWER COLOUR

red

pink

yellow

greenish yellow

orange

white

Bud number per cluster	Habit	Position of buds	Fruit width	Bark
1	tree			
	mallee			
	shrub			
3	tree	terminal	>1.5 cm	
			0.5–1.5 cm	
		axillary	>1.5 cm	rough
				smooth
			0.5–1.5 cm	rough
				smooth
			<0.5 cm	
	mallee	terminal		
		axillary	>1.5 cm	rough
				smooth
				minniritchi
			0.5–1.5 cm	rough
				smooth
			<0.5cm	
	shrub			
7	tree	terminal	>1.5 cm	rough
			0.5–1.5 cm	smooth

Crown colour	Concol/ discol	Other features in text	Species number
			93, 94, 121
			24, 80, 92, 93, 94 113,
			113
			7, 13–15
			9, 18, 131, 133
			28 , 45, 168
			27, 46, 58, 63, 97, 121
grey etc.			45, 114, 115, 119
green etc.			1, 30, 45, 107, 125, 156
grey etc.			102, 114, 115, 116, 117, 118, 181
green etc.		compound inflor-escences in axils	3, 22, 23, 35
		simple inflor-escences in axils	46, 50, 58, 97, 113, 116, 125, 156, 181
			35, 102, 113
			24, 133, 136
			28, 82–84, 168
grey etc.			26, 80, 81
green etc.			46, 80, 81,197, 198, 199
			87
			24, 30, 107, 114
grey etc.			24, 114, 118
green etc.			29, 46, 50, 55
			24
			113, 114, 117, 118, 181
	discolorous		5–8
	concolorous		12–16, 20
			11
grey etc.			133, 138, 140, 141, 149-151
green etc.	discolorous		4, 17, 18, 33, 153, 164
	concolorous	rough bark on trunk only	10, 11, 18, 47, 139, 142, 145, 146, 147
		rough bark at least to major limbs	9, 19, 21, 130, 131, 132, 148, 149, 150, 152, 164

Bud number per cluster	Habit	Position of buds	Fruit width	Bark
			<0.5 cm	
		axillary	>1.5 cm	rough
				smooth
				minniritchi
			0.5–1.5 cm	rough
				smooth
				minniritchi
7	tree	axillary	<0.5 cm	rough
				smooth

Crown colour	Concol/ discol	Other features in text	Species number
grey etc.			134, 140, 150, 151
green etc.	discolorous		157–159, 164
	concolorous		47, 132, 135, 139, 150, 152, 164
	discolorous		31, 43
	concolorous		28, 36, 167, 168,169, 191
			63
			85
grey etc.			96, 108, 127, 154, 155, 193
green etc.	discolorous	rough bark on trunk only	39, 40, 163
		rough bark at least to major limbs	41, 43, 120, 160, 162–164, 192, 193
	concolorous	rough bark on trunk only	1, 2, 47, 53, 60, 69, 107, 109, 122, 128, 155, 170, 171, 182, 183, 186, 187
		rough bark at least to major limbs	36, 47, 52, 73, 108, 124, 154, 164, 165, 167, 187
grey etc.			25, 98, 99, 100, 102, 103, 105, 155, 180, 181
green etc.	discolorous		37, 38, 44
	concolorous	operculum longer than base	51, 56, 57, 59, 61, 78, 103, 104
		short opercula	3, 35, 49, 54, 76, 99, 101, 106, 109, 112, 122, 123, 128, 155, 181, 190
			85
grey etc.			110, 126, 127
green etc.	discolorous		39, 40, 126, 160, 162, 164
	concolorous		47, 69, 122, 164
grey etc.			100, 102, 111
green etc.	discolorous		38
	concolorous		35, 57, 76, 104, 106, 112, 122, 144

Bud number per cluster	Habit	Position of buds	Fruit width	Bark
		bud clusters on leafless branchlets		
7	mallee	terminal		
		axillary	>1.5 cm	
			0.5–1.5 cm	rough
				smooth
				minniritchi
			<0.5 cm	
	shrub			
>7	tree	terminal	>1.5 cm	
			0.5–1.5 cm	
			<0.5 cm	
>7	tree	axillary	>1.5 cm	
			0.5–1.5 cm	rough

Crown colour	Concol/ discol	Other features in text	Species number
			34
grey etc.			133, 134, 138, 151
green etc.	discolorous		5, 161
	concolorous		132, 136, 137, 143, 146
			28, 85, 168, 196
grey etc.			96, 108, 193
green etc.		steeply ascending disc	107, 108
		disc not steeply ascending	52, 72, 73, 77, 91, 95, 109, 128, 193
grey etc.			48, 79, 90, 98, 180
green etc.		leaves up to 1 cm wide	57, 66, 70, 75, 174, 200
		leaves >1 cm wide	49, 51, 54, 55, 59, 70, 71, 74, 78, 89, 91, 95, 109, 128, 129, 194
			85, 86
			57, 66, 77, 88, 89, 143
			180, 181, 194
			13, 16
grey etc.			140, 149, 150
green etc.	discolorous		164
	concolorous		9, 47, 130, 131, 146, 148–150, 164
			47, 135, 140, 150, 164
			43, 64 (fused cluster), 167, (single fruit)
	discolorous		39–43, 160, 162, 164, 193
	concolorous	rough bark on trunk only	2, 47, 60, 62, 127, 170, 175, 176, 182, 186, 187, 189
		rough bark at least to major limbs	32, 73, 164, 165, 166, 167, 173, 177, 178, 187, 188

Bud number per cluster	Habit	Position of buds	Fruit width	Bark
				smooth
			<0.5 cm	rough
				smooth
		paired axillary		
		bud clusters on leafless branchlets		
>7	mallee	terminal		
		axillary	>1.5 cm	
			0.5–1.5 cm	rough
				smooth
			<0.5 cm	
	shrub			

Crown colour	Concol/ discol	Other features in text	Species number
grey etc.			65, 67, 100, 103, 105, 180, 181
green etc.		operculum longer than base	59, 65, 78, 103, 104
		short operculum	38, 76, 106, 179, 181, 185, 190
grey etc.			126, 127
green etc.	discolorous		39, 40, 126, 160, 162, 164
	concolorous		47, 164, 175, 176
grey etc.			65, 100
green etc.	discolorous		38
	concolorous		65, 76, 104, 106
			171, 172
			34
			146
			64 (fused cluster), 195
			32, 62, 73, 77, 193
grey etc.			90, 180
green etc.			59, 66, 68, 71, 74, 78, 184, 194, 200
			66, 77, 88
			180, 181, 194

1

Eucalyptus tessellaris

CARBEEN, MORETON BAY ASH

DESCRIPTION Small to tall erect tree. **Bark** rough, tessellated, dark grey on lower part of trunk with sharp change to smooth, grey or white above. **Juvenile leaves** shortly stalked, opposite for few pairs, then alternate, elliptical to lanceolate, to 24 x 5 cm. **Adult leaves** stalked, subopposite to alternate, narrow-lanceolate to linear, to 20 x 2 cm, concolorous, dull, green. **Bud clusters** on elongated leafless shoots in axils of leaves, in 3s or 7s. **Buds** stalked, ovoid to pear-shaped, to 0.7 x 0.5 cm; operculum shallowly hemispherical. **Flowers** white, November to February. **Fruit** stalked, urn-shaped to cylindrical, thin-walled, to 1.2 x 0.9 cm.

DISTRIBUTION Widely distributed in eastern half of Queensland including Torres Strait Islands, extending into north-eastern New South Wales. Usually on plains. Also in New Guinea.

NOTES Belongs to the ghost gums, which consist of about 20 species. Notable for its erect form, distinct basal rough bark, elongated floral shoots, and thin-walled fruits which shed their seed soon after formation.

Eucalyptus confertiflora

BROAD-LEAVED CARBEEN

DESCRIPTION Small to medium-sized tree. **Bark** rough, tessellated, dark grey on lower part of trunk, smooth, white above. **Juvenile leaves** shortly stalked, opposite for many pairs. **Mature crown** of scabrid late juvenile leaves, to 30 x 18 cm. New growth purple or claret-coloured. **Bud clusters** in dense compound masses on leafless branchlets usually inside current leaves, each up to 11-flowered. **Buds** on long slender stalks to 3 cm long. **Buds** pear-shaped, to 0.6 x 0.6 cm; operculum shallowly hemispherical. **Flowers** white, September to November. **Fruit** on long slender stalks, cup-shaped to cylindrical, thin-walled, to 1.2 x 1 cm.

DISTRIBUTION Widely distributed across northern Australia. Usually on plains in open woodlands.

NOTES Belongs to the ghost gums, which consist of about 20 species. Notable for its rough basal bark, usually deciduous crown in dry season, colourful new leaves, and long stalked buds and fruits.

Eucalyptus aparrerinja

GHOST GUM

DESCRIPTION Small to medium-sized tree. **Bark** smooth, powdery, white to cream. **Juvenile leaves** stalked, sub-opposite to alternate, ovate to broad-lanceolate, to 18 x 6 cm. **Adult leaves** stalked, lanceolate to narrow-lanceolate, to 16 x 2 cm, concolorous, glossy green. **Buds** densely clustered on short shoots in leaf axils, in 3s or 7s. **Buds** stalked, ovoid to pear-shaped, to 0.6 x 0.5 cm; operculum hemispherical to obtusely conical. **Flowers** white, December to February. **Fruit** stalked, cup-shaped to cylindrical, thin-walled, to 1.3 x 0.8 cm.

DISTRIBUTION Across central Australia from far eastern Western Australia to south-western Queensland.

NOTES The most famous of the ghost gums, it appears in many works of Albert Namatjira. It has been called *E. papuana* for many years, but this name should be reserved for a related species in Papua New Guinea and possibly on Cape York Peninsula. It now correctly assumes the Aboriginal name.

Eucalyptus gummifera

RED BLOODWOOD

DESCRIPTION Small to tall tree. **Bark** rough over trunk and branches, tessellated, brown. **Juvenile leaves** stalked, hairy, some peltate. **Adult leaves** stalked, alternate, lanceolate to broad-lanceolate, to 14 x 3.5 cm, discolorous, green. **Bud** clusters at leafless ends of branchlets, in 7s. **Buds** stalked, club-shaped, to 1.1 x 0.7 cm; operculum beaked. **Flowers** white, January to April. **Fruit** stalked, urn-shaped, to 2 x 1.4 cm.

DISTRIBUTION Largely coastal, from far eastern Victoria (e.g. Mallacoota) through New South Wales almost to Maryborough in south-eastern Queensland. Also inland from Toowoomba to Mt Walsh National Park. Sites vary from tall open forest to poor, sandy coastal heathlands.

NOTES This species belongs to the large bloodwood group, which occurs in all mainland States although greatest in numbers of species in the tropical north. Its unwinged seeds link it with two other southern species of the bloodwood group in far south-west Western Australia.

5

Eucalyptus calophylla

MARRI

DESCRIPTION Mallee to tall tree. **Bark** rough over trunk and branches, tessellated, grey-brown to dark brown. **Juvenile leaves** stalked, ovate, to 12 x 6 cm, some peltate. **Adult leaves** stalked, broad-lanceolate, to 15 x 5 cm, discolorous, green, with oil glands. **Bud** clusters at leafless ends of branchlets, in 7s. **Buds** stalked, pear-shaped, to 1.4 x 0.9 cm; operculum hemispherical to beaked. **Flowers** white or rarely pink, December to April. **Fruit** stalked, urn-shaped, to 5 x 3.8 cm.

DISTRIBUTION South-western Western Australia from Mt Lesueur area and New Norcia in the north to east of Albany, with a small disjunct occurrence near Ellendale Pool on the Greenough River. In hills and plains, usually on infertile sites, often on lateritic sandy gravels.

NOTES Marri is one of the four southern bloodwoods and shares with *E. gummifera* of south-eastern Australia the unwinged seeds and poor coastal sandy habitat.

Eucalyptus ficifolia

RED FLOWERING GUM

DESCRIPTION Small tree. **Bark** rough over trunk and branches, fibrous and furrowed or somewhat tessellated. **Juvenile leaves** stalked, ovate to broad-lanceolate, some peltate, to 12 x 6 cm. **Adult leaves** stalked, broad-lanceolate, to 14 x 6 cm, discolorous, green, without oil glands. **Bud** clusters at leafless ends of branchlets, in 7s. **Buds** on long stalks, to 1.6 x 0.8 cm; operculum conical to beaked. **Flowers** orange, red or pink, January to April. **Fruit** stalked, barrel-shaped, to 4.2 x 3.1 cm.

DISTRIBUTION In far south-western Western Australia, restricted to a small area east of Mt Frankland and Walpole. On infertile coastal sandy plains.

NOTES One of the four southern bloodwoods and distinguished from the others by the winged seeds and spectacular coloured flowers. A favourite ornamental in southern Australia.

7

Eucalyptus abergiana

RANGE BLOODWOOD

DESCRIPTION Small tree. **Bark** rough over trunk and branches, tessellated, grey-brown. **Juvenile leaves** stalked, lanceolate to ovate, hairy, to 15 x 5 cm. **Adult leaves** stalked, broad-lanceolate, to 16 x 5 cm, discolorous, green, with oil glands. **Bud** clusters at leafless ends of branchlets, in 3s or 7s. **Buds** without stalks, ovoid, sometimes white, waxy, to 1.5 x 1 cm; operculum conical, warty. **Flowers** white, August to November. **Fruit** without stalks, cup-shaped with thick rim, to 3 x 2.5 cm.

DISTRIBUTION Northern Queensland, in ranges from Paluma to west of Cairns.

NOTES *E. abergiana* resembles no other bloodwood with its large thick-rimmed fruit without stalks.

Eucalyptus ptychocarpa

SPRING BLOODWOOD

DESCRIPTION Small to medium-sized tree. **Bark** rough over trunk and branches, tessellated, brownish grey. **Juvenile leaves** stalked, elliptical to broad-lanceolate, to 40 x 15 cm. **Adult leaves** stalked, broad-lanceolate, to 25 x 7 cm, discolorous, green. **Bud** clusters at leafless ends of branchlets, in 7s. **Buds** stalked, club-shaped to ovoid, usually ribbed, to 3 x 1.8 cm; operculum hemispherical to broadly conical. **Flowers** red, pink, apricot or rarely white, September to March. **Fruit** stalked, +/– urn-shaped, ribbed, to 6 x 4 cm.

DISTRIBUTION Northern Australia, from the Kimberley east to far north-western Queensland. Usually along creeks and in low-lying places.

NOTES A bloodwood recognised by the large leaves, large ribbed buds and fruit, and spectacular coloured flowers.

Eucalyptus nesophila

MELVILLE ISLAND BLOODWOOD

DESCRIPTION Small to occasionally tall tree. **Bark** rough over trunk and branches, tessellated, grey-brown. **Juvenile leaves** stalked, ovate to heart-shaped, sometimes peltate, hairy to scabrid, to 15 x 9 cm. **Adult leaves** stalked, lanceolate to narrow-lanceolate, to 18 x 1.8 cm, concolorous, green. **Bud** clusters at leafless ends of branchlets, in 3s, 7s or more. **Buds** stalked, club-shaped to pear-shaped, to 0.7 x 0.5 cm; operculum hemispherical with a point. **Flowers** white, June to August. **Fruit** stalked, elongated and urn-shaped, to 1.5 x 0.9 cm.

DISTRIBUTION Northern Australia, in the north Kimberley of Western Australia, Melville Island and Cobourg Peninsula of the Northern Territory, and northern Cape York Peninsula and islands of Torres Strait. On low hills and plains.

NOTES One of the winged-seeded bloodwoods, it is recognised in the field by the fully rough bark, cordate juvenile leaves, glossy green adult leaves, and delicate elongated fruit.

Eucalyptus bleeseri

SMOOTH-STEMMED BLOODWOOD

DESCRIPTION Small to medium-sized tree. **Bark** rough over lower part of trunk, loose, thin and flaky, rest of trunk smooth, creamy white. **Juvenile leaves** stalked, ovate, peltate, hairy, to 6 x 4 cm. **Adult leaves** stalked, lanceolate, to 18 x 2.5 cm, concolorous, glossy, bright green. **Bud** clusters at leafless ends of branchlets; in 7s. **Buds** stalked, ovoid, to 1.2 x 0.6 cm; operculum conical. **Flowers** creamy white, March to June. **Fruit** stalked, ovoid to urn-shaped, to 2.2 x 1.5 cm.

DISTRIBUTION The northern Kimberley of Western Australia and the Top End of the Northern Territory. On plains and low hills on sandstone and laterites.

NOTES A winged-seeded bloodwood, recognised by the scaly bark and very glossy green leaves. It is notable for the sudden change from the juvenile leaf stage to the smooth-leaved intermediate leaves on saplings about 1 m tall.

Eucalyptus dichromophloia

SMALL-FRUITED BLOODWOOD

DESCRIPTION Small to medium-sized tree. **Bark** varying from completely smooth to partly rough, flaky on lower trunk. **Juvenile leaves** stalked, elliptical, to 17 x 6.5 cm. **Adult leaves** stalked, narrow-lanceolate to lanceolate, to 20 x 2.5 cm, concolorous, green, usually glandular. **Bud** clusters at leafless ends of branchlets, in 7s. **Buds** stalked, pear-shaped, to 0.5 x 0.4 cm; operculum conical to hemispherical. **Flowers** creamy white, March–May. **Fruit** stalked, urn-shaped, to 1.3 x 1 cm.

DISTRIBUTION Northern Australia from the central Kimberley through the Top End of the Northern Territory to far north-western Queensland. On plains and low stony hills.

NOTES A winged-seeded bloodwood. A very variable species in the bark characters and size of fruit. Forms on the Kimberley plateau are notable for the silvery new leaf growth at the outside of the crown. These have been segregated as *E. drysdalensis*, but the distinction is slight.

Eucalyptus terminalis

DESERT BLOODWOOD

DESCRIPTION Small to medium-sized tree. **Bark** variable, rough on part or whole of trunk, tessellated, mottled brown, orange and cream. **Juvenile leaves** stalked, lanceolate, to 15 x 3 cm. **Adult leaves** stalked, lanceolate, to 15 x 2.5 cm, concolorous, green, usually without oil glands. **Buds** in clusters at leafless ends of branchlets, in 7s. **Buds** stalked, ovoid to club-shaped, cream-coloured and scurfy (rough-surfaced), to 1 x 0.7 cm; operculum hemispherical to shallowly conical. **Flowers** creamy white, March to August. **Fruit** stalked, urn-shaped, to 2.4 x 2 cm.

DISTRIBUTION Widely distributed across northern Australia, particularly arid areas, extending to south of the River Darling between Cobar and Louth.

NOTES The most widely occurring bloodwood. It is a winged-seeded bloodwood which is particularly conspicuous in flower, when the large creamy flower clusters can be seen from a distance. As with most desert bloodwoods, the leaves lack an intramarginal vein and have no visible oil glands.

13

Eucalyptus chippendalei

SAND-DUNE BLOODWOOD

DESCRIPTION Small to medium-sized tree. **Bark** rough over part or most of trunk, tessellated, thin, flaking, yellow-brown. **Juvenile leaves** stalked, lanceolate, to 8 x 3 cm. **Adult leaves** stalked, lanceolate, to 11 x 2.5 cm, concolorous, green, without oil glands. **Bud** clusters at leafless ends of branchlets, in 3s or 7s, rarely 11s. **Buds** stalked, ovoid to pear-shaped, to 0.9 x 0.9 cm; operculum hemispherical to shallowly conical. **Flowers** creamy white, summer. **Fruit** stalked, +/– spherical but cut off at the top, to 2.6 x 2.5 cm.

DISTRIBUTION Scattered in deserts of central eastern Western Australia, extending into the Northern Territory near Uluru. Always on top or side of desert sand-dunes.

NOTES A winged-seeded bloodwood, it is usually conspicuous for its peculiar habitat and the fruits usually lacking a neck.

Eucalyptus foelscheana

BROAD-LEAVED BLOODWOOD

DESCRIPTION Small to medium-sized tree. **Bark** rough over part or most of trunk, thin, loose, flaky, red-brown, orange or creamy white. **Juvenile leaves** stalked, ovate to elliptical, some peltate, slightly hairy, to 35 x 25 cm. **Adult leaves** stalked, broadly ovate to almost round, to 20 x 12 cm, concolorous, green. **Bud** clusters at leafless ends of branchlets, in 3s or 7s. **Buds** stalked, club-shaped to pear-shaped, to 1 x 0.8 cm; operculum hemispherical. **Flowers** creamy white, April to October. **Fruit** stalked, urn-shaped, to 2.5 x 2 cm.

DISTRIBUTION Top End of the Northern Territory from Tindal northwards. Open woodland on plains, slopes and low hills.

NOTES A winged-seeded bloodwood, it is recognised by the loose rough bark and very large leaves.

Eucalyptus setosa

ROUGH-LEAVED BLOODWOOD

DESCRIPTION Small tree, often with several upright stems. **Bark** rough over trunk and branches, tessellated, grey-brown. **Whole crown** of juvenile leaves, without stalks, opposite, ovate, scabrid, to 7 x 4.5 cm, concolorous, green. **Buds** in clusters at leafless ends of branchlets, in 3s or 7s. **Buds** stalked, club-shaped, scabrid/hairy, to 0.8 x 0.7 cm; operculum beaked. **Flowers** white, November to December. **Fruit** stalked, slightly urn-shaped, rough-surfaced, to 2.3 x 2 cm.

DISTRIBUTION Central Northern Territory, eastwards to central and parts of northern Queensland. On plains.

NOTES A winged-seeded bloodwood, it is one of a group of species recognised by the crown of ovate, scabrid leaves and scabrid or hairy buds and fruits.

Eucalyptus zygophylla

BROOME BLOODWOOD

DESCRIPTION Small straggly tree. **Bark** rough over trunk and branches, coarsely tessellated, grey. **Whole crown** of juvenile leaves, without stalks, broad-lanceolate, thick, to 15 x 5 cm, concolorous, green. **Bud** clusters at leafless ends of branchlets, in 7s or 11s. **Buds** on short thick stalks or lacking stalks, pear-shaped, to 1.2 x 1.5 cm; operculum slightly beaked. **Flowers** creamy white, January to March. **Fruit** shortly stalked or lacking stalks, ovoid cut off at the top, to 3.5 x 3 cm, with thick rim.

DISTRIBUTION Western Australia fom the Exmouth Gulf region and Nanutarra north-eastwards to Broome, Derby and Fitzroy Crossing. On plains.

NOTES A winged-seeded bloodwood, easily recognised by the large opposite leaves and the large, thick-rimmed fruit. Not to be confused with *E. cadophora* (the formerly known *E. perfoliata*), which has similarly large opposite leaves but which are joined at their bases across the stem.

Eucalyptus trachyphloia

BROWN BLOODWOOD

DESCRIPTION Medium-sized to tall tree. **Bark** rough on trunk and larger branches, flaky and irregularly tessellated, yellow-brown. **Juvenile leaves** stalked, opposite, elliptical to lanceolate, some peltate, to 10 x 1.5 cm, early leaves hairy, later leaves glossy green and scabrid below. **Adult leaves** stalked, narrow-lanceolate to lanceolate, to 15 x 1.5 cm, discolorous, green. **Bud** clusters at leafless ends of branchlets, in 7s. **Buds** stalked, pear-shaped, to 0.5 x 0.3 cm; operculum beaked. **Flowers** white to cream, January to March. **Fruit** stalked, urn-shaped or barrel-shaped, to 1 x 0.8 cm.

DISTRIBUTION Eastern Queensland from the Newcastle Range and Atherton Tableland southwards to north-eastern New South Wales. On plains and hills.

NOTES Somewhat like a yellow bloodwood in the bark and has similar seeds, i.e. red-brown, shiny and unwinged. The buds and fruit are delicate compared with most typical bloodwoods.

18

Eucalyptus torelliana

CADAGHI

DESCRIPTION Medium-sized to tall tree. **Bark** rough on lower part of trunk, flaky to tessellated, grey to black. **Juvenile leaves** stalked, ovate, peltate, scabrid, to 22 x 14 cm. **Crown** often composed of juvenile and intermediate leaves. **Adult leaves** stalked, broad-lanceolate to lanceolate, 14 x 3.5 cm, slightly discolorous, green. **Bud** clusters at leafless ends of branchlets, in 3s, rarely 7s. **Buds** shortly stalked or lacking stalks, ovoid, to 0.9 x 0.6 cm; operculum hemispherical to broadly conical. **Flowers** white, September to October. **Fruit** without stalks, urn-shaped to almost spherical, to 1.4 x 1.4 cm.

DISTRIBUTION Restricted to northern Queensland from west of Ingham to south of Cooktown. Coastal foothills and ranges, sometimes into rainforest.

19

Eucalyptus peltata

YELLOWJACKET

DESCRIPTION Small to medium-sized tree. **Bark** rough, loose, flaky on trunk and large branches, golden yellow. **Juvenile leaves** shortly stalked, ovate, peltate, scabrid, to 21 x 12 cm. **Mature crown** of smaller juvenile leaves, becoming smooth, to 11 x 8 cm, concolorous, green. **Bud** clusters at leafless ends of branchlets, in 7s. **Buds** +/– without stalks, ovoid, to 0.8 x 0.6 cm; operculum conical often different colour to base of bud. **Flowers** white, January to February. **Fruit** without stalks, ovoid to almost spherical and cut off at the top, to 1.1 x 1 cm.

DISTRIBUTION Northern Queensland from the Newcastle range to south of Homestead. On hills and plains.

NOTES A yellow bloodwood, this species is easy to recognise by the yellow flaky bark and the broad, scabrid, often peltate leaves. Yellow bloodwoods differ from the typical (red) bloodwoods most prominently by the unwinged seeds and the early loss of the outer operculum.

Eucalyptus watsoniana

YELLOWJACKET

DESCRIPTION Small to medium-sized tree. **Bark** rough, loose, flaky on trunk and large branches, bright orange-yellow to brownish yellow. **Juvenile leaves** stalked, ovate to broad-lanceolate, to 22 x 10 cm. **Adult leaves** stalked, lanceolate to narrow-lanceolate, to 18 x 4.5 cm, concolorous, dull, light green to grey-green. **Bud** clusters at leafless ends of branchlets, in 7s. **Buds** with funnel-shaped base and turban-shaped operculum wider than the base, to 2 x 1.6 cm. **Flowers** creamy white, June to September. **Fruit** stalked, ovoid to urn-shaped, to 3 x 2.5 cm.

DISTRIBUTION South-eastern Queensland from Wigton to Robinson Gorge including the Isla Gorge area. Favours sandstone ridges and shallow sandy soils.

NOTES This species is distinguished from other yellow bloodwoods by having the largest buds and fruits. Crowns are conspicuous by the large leaves.

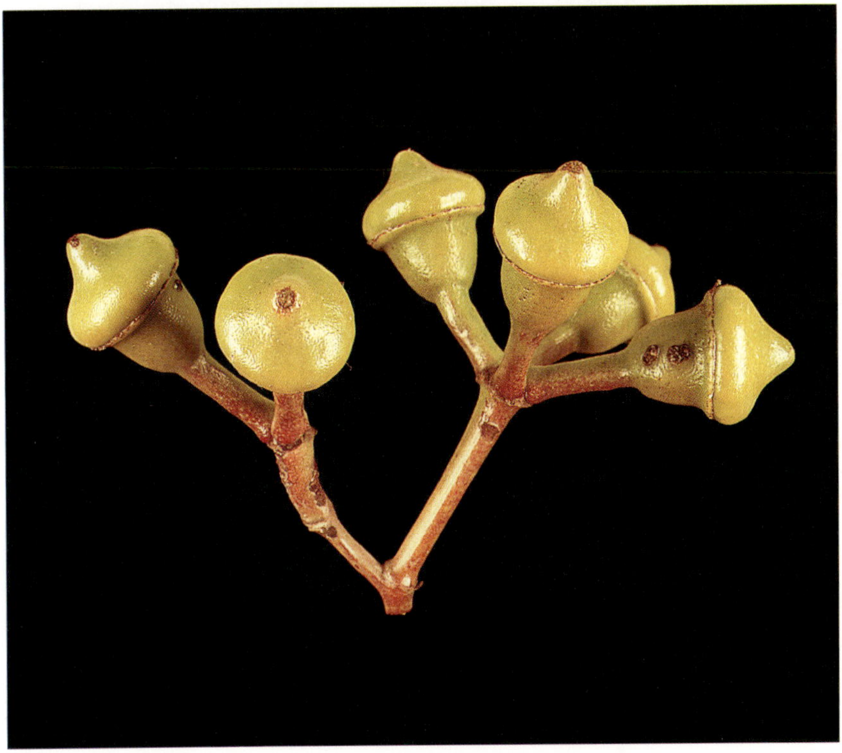

Eucalyptus eximia

YELLOW BLOODWOOD

DESCRIPTION Small to medium-sized tree. **Bark** rough, loose, flaky on trunk and large branches. **Juvenile leaves** stalked, elliptical to broad-lanceolate, many peltate, to 16 x 9 cm. **Adult leaves** stalked, curved, to 21 x 2.8 cm, concolorous, blue-green to green. **Bud** clusters at leafless ends of branchlets, in 7s. **Buds** lacking or on short stalks, club-shaped, to 1.2 x 0.9 cm; operculum beaked. **Flowers** creamy white, September to November. **Fruit** lacking or on short stalks, ovoid, to 1.8 x 1.5 cm.

DISTRIBUTION New South Wales from Pokolbin to Nowra on shallow sandy soils of the Hawkesbury Sandstone.

NOTES The only yellow bloodwood in New South Wales. It is notable for its large curved leaves and sandstone habitat.

Eucalyptus citriodora

LEMON-SCENTED GUM

DESCRIPTION Medium-sized to tall tree. **Bark** smooth, white, coppery or pink, often powdery, shedding in large curling flakes. **All leaves** lemon-scented. **Juvenile leaves** stalked, ovate to broad-lanceolate, peltate, hairy, to 21 x 8 cm. **Adult leaves** stalked, narrow-lanceolate, to 16 x 1.8 cm, concolorous, glossy green. **Bud clusters** in 3s on short leafless shoots in axils of leaves. **Buds** stalked, club-shaped, to 1 x 0.6 cm; operculum conical to slightly beaked. **Flowers** creamy white, June to November. **Fruit** stalked, urn-shaped, to 1.5 x 1 cm.

DISTRIBUTION Eastern Queensland from Maryborough to the Windsor Tableland. On dry ridges and uplands.

NOTES While endemic to subtropical areas, this species is a favourite ornamental in southern Australia, notable for the smooth bark and fresh green, lemon-scented leaves. Only two eucalypts are lemon-scented, this and *E. staigeriana* of Cape York Peninsula.

23

Eucalyptus maculata

SPOTTED GUM

DESCRIPTION Medium-sized to tall tree. **Bark** smooth, shedding in small, irregular dark grey flakes, becoming creamy white. **Juvenile leaves** stalked, elliptical to ovate, peltate, to 23 x 9.5 cm. **Adult leaves** stalked, lanceolate, to 18 x 2.5 cm, concolorous, slightly glossy, green. **Bud clusters** in 3s on short leafless shoots in axils of leaves. **Buds** stalked +/– club-shaped, to 1.1 x 0.7 cm; operculum conical to beaked. **Flowers** white to cream, May to September. **Fruit** stalked, barrel-shaped, to 1.4 x 1.1 cm.
DISTRIBUTION South-eastern Australia from Carnarvon Gorge in Queensland along the eastern coast to far eastern Victoria, north-west of Orbost.
NOTES A beautiful ornamental straight-trunked tree easily recognised among the largely rough-barked trees of the coastal forests.

Eucalyptus gamophylla

TWIN LEAF MALLEE

DESCRIPTION Erect mallee over most of its occurrences, but almost prostrate in central northern Western Australia. **Bark** smooth or rough. **Pith** of branchlets glandular. **Crown** composed of juvenile leaves in pairs joined around the stem, together to 12 x 3.5 cm, concolorous, blue-grey to waxy. **Bud clusters** at ends of branchlets or in leaf axils, buds single or in 3s. **Buds** stalked, club-shaped, to 0.6 x 0.4 cm; operculum hemispherical. **Flowers** white, October to March. **Fruit** stalked, +/– cylindrical to funnel-shaped, usually square at the opening with small projections, to 1.6 x 0.7 cm.

DISTRIBUTION Widespread in central arid Australia from the Pilbara of Western Australia eastwards to far western Queensland.

NOTES Belongs to the small *Eudesmia* group which occurs in WA, SA, NT, Qld and NSW and includes the following eight species. They vary greatly in form, from desert mallee to forest tree. This species is a mallee of arid zones easily recognised by the bluish, joined (connate) leaves invariably present. Some plants may be more advanced and produce a high proportion of narrower, unjoined leaves, which are characteristic of the closely related *E. odontocarpa*.

Eucalyptus gongylocarpa

MARBLE GUM

DESCRIPTION Small to medium-sized tree. **Bark** smooth, white, or with loosely attached flakes. **Pith** of branchlets glandular. **Leaves**, buds, fruits, stems white, waxy. **Juvenile leaves** lacking a stalk, opposite, round to ovate, to 5 x 3.5 cm, white, waxy. **Adult leaves** shortly stalked, opposite to sub-opposite, oblong-elliptical to lanceolate, to 7.5 x 1.7 cm, concolorous, light green, becoming grey to waxy with age. **Buds** in leaf axils, in 7s, stalked, club-shaped, to 0.5 x 0.4 cm; operculum hemispherical. **Flowers** white, January to February. **Fruit** stalked, almost spherical, to 1.1 x 1.2 cm.

DISTRIBUTION Central Western Australia extending into the Northern Territory near Lake Amadeus and in the Great Victoria Desert of South Australia. On sandy plains and low sandy hills.

NOTES One of the finest of the desert gums, with its gleaming white bark and soft grey foliage.

26

Eucalyptus tetragona

TALLERACK

DESCRIPTION Mallee. **Bark** smooth. **Leaves**, buds, fruits, stems white waxy. **Pith** of branchlets glandular. **Crown** of mature mallee composed wholly of juvenile leaves, opposite, elliptical, to 16 x 8 cm, glaucous. **Buds** in axils of leaves, in 3s, stalked, cylindrical to pear-shaped, square in section, to 1 x 0.8 cm; operculum flattened-hemispherical. **Flowers** white, December to February. **Fruit** stalked, cuboid, square in section, to 2.7 x 2.3 cm.

DISTRIBUTION Widespread on the sandplains of south-western Western Australia from Eneabba in the north to east of Esperance in the south.

NOTES Easily recognised by the spindly mallee form, large, white, waxy, opposite leaves, and when in flower, by the stamens seen clustered in four groups.

Eucalyptus erythrocorys

ILLYARRIE

DESCRIPTION Small tree usually of poor form. **Bark** smooth, pale yellow to white, often with loose flakes of dead bark. **Pith** of branchlets glandular. **Juvenile leaves** petiolate, elliptical to ovate, to 10 x 6 cm, grey, densely hairy. **Adult leaves** petiolate, opposite, sickle-shaped, to 18 x 2.7 cm, concolorous, glossy, green. **Buds** in axils of leaves, in 3s, on strongly flattened stalks, to 3 x 3 cm, ribbed at base, with bright red 4-lobed operculum. **Flowers** (stamens) yellow, February to April. **Fruit** shortly stalked, +/– 4-sided, ribbed, to 4 x 5 cm; disc broad, level or partly sunken.

DISTRIBUTION Western Australia, from Cockleshell Gully to north-east of Dongara usually on undulating limestony low rises.

NOTES Illyarrie is the most spectacular species of the *Eudesmia* group and is a favourite ornamental in southern Australia.

28

Eucalyptus roycei

SHARK BAY MALLEE

DESCRIPTION Mallee or small tree. **Bark** rough at the base, grey-brown, smooth above. **Juvenile leaves** stalked, opposite, hairy, ovate, to 9 x 5 cm. **Adult leaves** stalked, subopposite to alternate, lanceolate or sickle-shaped, to 15 x 3 cm, concolorous, dull, light green to grey-green. **Buds** in axils of leaves, in 7s rarely 3s, shortly stalked, oblong and square in section, waxy, to 2 x 1.3 cm; operculum pyramidal to flattened-hemispherical. **Flowers** pale yellow, January to March. **Fruit** shortly stalked, +/– cuboid, to 3 x 2.7 cm.

DISTRIBUTION Western Australia in the area south of Shark Bay and west of Wannoo. Usually on low sandy rises.

NOTES A small tree or mallee, which has spectacular dense flower clusters and grows well in cultivation in Perth.

Eucalyptus ebbanoensis

SANDPLAIN MALLEE

DESCRIPTION Mallee. **Bark** rough, flaky at base or smooth, grey to greenish or coppery. **Juvenile leaves** stalked, ovate, hairy, to 11 x 3.5 cm. **Adult leaves** stalked, lanceolate or sickle-shaped, to 14.5 x 2.5 cm, concolorous, green, dull over most of range (subsp. *ebbanoensis*) but glossy in the Moresby Range (subsp. *photina*). **Buds** in axils of leaves, in 3s, stalked, ovoid to club-shaped, to 0.8 x 0.6 cm; operculum flattened-hemispherical. **Flowers** white, September to March. **Fruit** stalked, cup-shaped, to 1.1 x 1.1 cm.

DISTRIBUTION Western Australia, widespread from the Moresby Range north of Geraldton to the western part of the Great Victoria Desert north-east of Kalgoorlie. Mostly on sandy plains but on rocky mesas in the Moresby Range.

NOTES A mallee, fairly easy to identify with its buds in 3s, flat, thick-rimmed fruit with 3 valves, and in flower by the stamens being clustered in four groups.

Eucalyptus lirata

KIMBERLEY YELLOWJACKET

DESCRIPTION Small tree or mallee. **Bark** rough to small branches, loose, papery, friable, yellow. **Juvenile leaves** stalked, opposite, broad-lanceolate, to 9 x 3.5 cm. **Adult leaves** stalked, mostly sickle-shaped, to 14 x 2 cm, concolorous, dull, blue-green to green. **Buds** in axils of leaves, in 3s, cylindrical to pear-shaped, to 1 x 0.4 cm; operculum hemispherical to obtusely conical. **Flowers** November to February. **Fruit** stalked, cup-shaped, to 1.3 x 0.9 cm.

DISTRIBUTION Central Kimberley of Western Australia, north and east of Gibb River. On plains and low stony hills.

NOTES One of two related northern yellowjackets, this species is distinguished from the related *E. similis* of Queensland by the buds in 3s and the almost concolorous leaves.

Eucalyptus miniata

DARWIN WOOLLYBUTT

DESCRIPTION Small to medium-sized tree. **Bark** rough on part or most of trunk, thick, firm or papery, dark brown or yellow-brown. **Juvenile leaves** stalked, ovate, to 12 x 5 cm, green, hairy. **Adult leaves** stalked, lanceolate to broad-lanceolate, to 16 x 3.5 cm, discolorous, light green. **Buds** in axils of leaves, in 7s, on stout stalks, club-shaped, ribbed, to 2.3 x 1.1 cm; operculum hemispherical to conical. **Flowers** orange, March to July. **Fruit** lacking a stalk, truncate-ovoid to slightly urn-shaped, ribbed, to 6 x 5 cm.

DISTRIBUTION Northern Australia, from the Kimberley across the Top End of the Northern Territory to the southern part of Cape York Peninsula, south to the Pentland area. On plains, low hills and tablelands, usually on poor soils.

NOTES A very common tree in northern Australia easily recognised by the brilliant orange flowers and large, ribbed fruit.

Eucalyptus phoenicea

SCARLET GUM

DESCRIPTION Small to medium-sized tree or mallee. **Bark** rough on trunk and larger branches, flaky, yellowish or yellowish grey. **Juvenile leaves** stalked, elliptical to ovate, to 12 x 6 cm, hairy. **Adult leaves** stalked, lanceolate, to 12 x 2 cm, concolorous, dull, light green to yellowish green. **Buds** in axils of leaves, to > 20-flowered, on long tapering stalks, club-shaped, to 1.8 x 0.7 cm; operculum hemispherical to conical. **Flowers** orange, April to August. **Fruit** stalked, cylindrical and contracted into a neck at the top, to 3 x 1.3 cm.

DISTRIBUTION From the east Kimberley of Western Australia to the Top End of the Northern Territory; also north-west of Cooktown in Cape York Peninsula. On scarps and stony sites.

NOTES A species easy to recognise by the brilliant orange flowers and yellowish, flaky bark. In the Kimberley the coloured flowers and smooth upper limbs distinguish it from the other yellowjacket, *E. lirata*.

Eucalyptus guilfoylei

YELLOW TINGLE

DESCRIPTION Medium-sized to tall tree. **Bark** rough on trunk and larger branches, fibrous, grey-brown. **Juvenile leaves** stalked, elliptical to ovate, to 13 x 7 cm. **Adult leaves** stalked, lanceolate, to 16 x 4 cm, discolorous, dull green. **Bud clusters** at leafless ends of branchlets, in 7s. **Buds** shortly stalked, oblong to club-shaped, to 0.9 x 0.5 cm; operculum hemispherical to obtusely conical. **Flowers** white, November to December. **Fruit** lacking or with short stalks, cupular, to 1 x 1 cm.

DISTRIBUTION Very restricted distribution in the far south-west of Western Australia, in the vicinity of Bow Bridge and Walpole.

NOTES One of the rarest eucalypts, this species is easy to recognise in its natural area of occurrence by the terminal bud clusters and rough but not flaky bark. It is closely related to no other eucalypt and retains many primitive botanical features, such as the terminal bud clusters, rough bark, discolorous leaves with wide-spreading venation, and the curious and rare sutures on the operculum, probably representing evidence of fused petals.

Eucalyptus cladocalyx

SUGAR GUM

DESCRIPTION Small to tall tree. **Bark** smooth, mottled orange, yellow and grey, over whole trunk although usually with partially detached fragments of dead bark. **Juvenile leaves** stalked, lanceolate, round, to 6 x 9 cm. **Adult leaves** stalked, lanceolate, to 15 x 2.4 cm, strongly discolorous, dark green on top side. **Bud clusters** on leafless parts of the branchlets inside the current leaves, i.e. the second-year growth, in 7s or more. **Buds** stalked, cylindrical, to 1 x 0.6 cm; operculum hemispherical. **Flowers** creamy white, January to February. **Fruit** stalked, barrel-shaped, ribbed, to 1.5 x 1 cm.

DISTRIBUTION South Australia, restricted to southern parts of the Flinders Range, Eyre Peninsula and Kangaroo Island.

NOTES A species not related closely to any other eucalypt, it is probably a tall forest relic species from times of wetter climate. It is easily recognised by the strongly discolorous leaves and the placement of the bud clusters on leafless branchlets inside the crown. Sugar gum is one of the most widely planted eucalypts in southern Australia.

Eucalyptus michaeliana

HILLGROVE GUM

DESCRIPTION Small to medium-sized tree. **Bark** smooth or with loose flakes of unshed dead bark, mottled white, grey, yellow or red. **Juvenile leaves** stalked, broad-lanceolate, to 15 x 4 cm. **Adult leaves** stalked, lanceolate, to 20 x 3 cm, concolorous, dull green. **Bud clusters** in groups of 3 (rarely 5) in axils of leaves, each 3- or 7-flowered. **Buds** stalked, ovoid, to 0.5 x 0.2 cm; operculum conical to hemispherical. **Flowers** white, rarely pink, August to November. **Fruit** stalked, cupular, to 0.5 x 0.5 cm.

DISTRIBUTION Restricted to Mt Barney National Park in Queensland and in New South Wales to the Wyong, Hillgrove and Enmore areas. Low hills.

NOTES A species not closely related to any other, it is the only one in south-eastern Australia, apart from the very different *E. tessellaris* and *E. maculata*, that have the compound bud clusters in the leaf axils.

Eucalyptus gomphocephala

TUART

DESCRIPTION Small to tall tree. **Bark** rough over trunk and branches, grey, fibrous. **Juvenile leaves** stalked, ovate, to 15 x 10 cm. **Branchlets** yellow. **Adult leaves** stalked, lanceolate or sickle-shaped, to 16 x 2.5 cm, concolorous, glossy green. **Buds** in axils of leaves, in 7s, on short, stout stalks, mushroom-shaped, to 2.3 x 1.5 cm; operculum hemispherical to obtusely conical. **Flowers** white, January to April. **Fruit** on short, stout stalks or stalk lacking, cup-shaped or bell-shaped (inverted), to 2.2 x 1.7 cm.

DISTRIBUTION South-west Western Australia, on coastal and subcoastal limestony plains and low dunes from Ludlow in the south to Jurien Bay in the north.

NOTES A species not closely related to any other, tuart is one of the three prominent rough-barked trees in its area of occurrence (the others being jarrah and marri). It is easily recognised by the bark and buds.

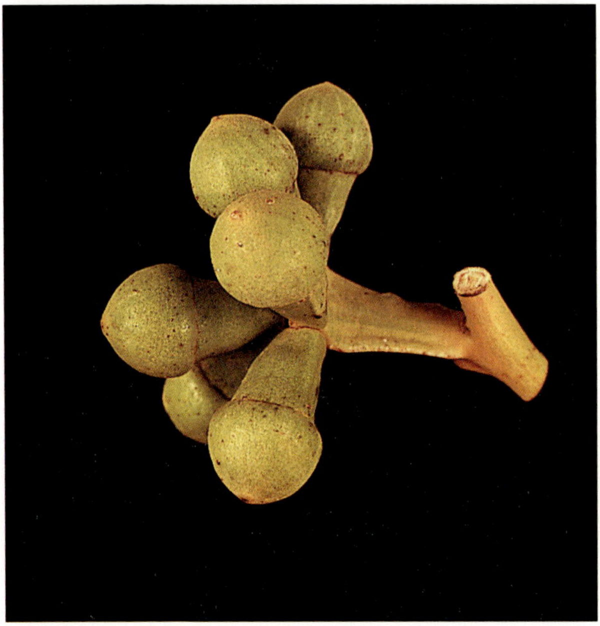

37

Eucalyptus diversicolor

KARRI

DESCRIPTION Medium-sized to very tall tree. **Bark** smooth, grey, orange, yellow or whitish. **Juvenile leaves** stalked, round to ovate, to 11 x 10 cm. **Adult leaves** stalked, lanceolate, to 12 x 3.2 cm, discolorous, dull green. **Buds** in axils of leaves, in 7s, stalked, club-shaped, to 1.4 x 0.7 cm; operculum conical. **Flowers** white, September to February. **Fruit** stalked, barrel-shaped, to 1.2 x 1 cm.

DISTRIBUTION The far south-west of Western Australia, in the high-rainfall areas from Boranup on the west coast east to Mount Manypeaks, including the Porongorup Range but not the Stirling Range.

NOTES One of the tallest trees in Australia, karri is a valuable hardwood timber and a famous tourist attraction because of its size and beautiful coloured bark in season.

Eucalyptus deanei

ROUND-LEAVED GUM

DESCRIPTION Medium-sized to very tall tree. **Bark** smooth, mottled, bright yellow on first exposure then creamy yellow, grey and brown. **Juvenile leaves** stalked, ovate to triangular, to 15 x 8.5 cm. **Adult leaves** stalked, lanceolate to broad-lanceolate, to 14 x 3.5 cm, discolorous, green. **Buds** in axils of leaves, in 7s to 11s, stalked, club-shaped or pear-shaped, to 0.6 x 0.4 cm; operculum hemispherical, conical or beaked. **Flowers** white, March to May. **Fruit** stalked, cup-shaped or bell-shaped (inverted), to 0.6 x 0.6 cm.

DISTRIBUTION Central New South Wales, from about Picton Lakes northwards to near Singleton.

NOTES Populations in northern New South Wales and into Queensland, previously called *E. deanei*, have been recently segregated as *E. brunnea,* although the differences between the two species are slight. The common name refers to the juvenile or intermediate leaf stage usually seen on the lower branches.

Eucalyptus grandis

FLOODED GUM, ROSE GUM

DESCRIPTION Medium-sized to tall tree. **Bark** smooth or rough at base, fibrous or flaky, grey to grey-brown. **Juvenile leaves** stalked, ovate, to 16 x 8.5 cm. **Adult leaves** stalked, lanceolate to broad-lanceolate, to 15 x 3 cm, discolorous, glossy dark green. **Buds** in axils of leaves, in 7s to 11s, stalked, pear-shaped or spindle-shaped, to 0.8 x 0.5 cm; operculum conical or beaked. **Flowers** white, April to August. **Fruit** shortly stalked or stalk lacking, funnel-shaped to cupular, 0.8 x 0.7 cm, valves incurved.

DISTRIBUTION Coastal areas and sub-coastal ranges from about Newcastle in New South Wales northwards to west of Daintree in northern Queensland. On flats and lower slopes.

NOTES A beautiful straight-trunked tree of east coast forests, much in demand overseas for its timber and fibre. It is most easily distinguished from Sydney blue gum by the incurved valves of the fruit. Buds and fruit may be white, waxy.

Eucalyptus saligna

SYDNEY BLUE GUM

DISTRIBUTION Coastal areas and subcoastal ranges from southern New South Wales northwards to south-eastern Queensland, where it extends further inland to the Blackdown and Consuelo Tablelands.

NOTES A tall tree valuable for its timber. It is most easily distinguished from flooded gum by the valves of the fruit, which are erect and usually curved outwards.

DESCRIPTION Medium-sized to tall tree. **Bark** rough at base or higher, thick, fibrous, brownish grey. **Juvenile leaves** stalked, ovate, to 12 x 5 cm. **Adult leaves** stalked, lanceolate to broad-lanceolate, to 16 x 3 cm, discolorous, green. **Buds** in axils of leaves, in 7s to 11s, stalked or lacking a stalk, narrowly pear-shaped or spindle-shaped, to 0.9 x 0.4 cm; operculum conical or beaked. **Flowers** white, May to July. **Fruit** shortly stalked or stalk lacking, funnel-shaped to cup-shaped, to 0.8 x 0.7 cm; valves erect.

Eucalyptus botryoides

BANGALAY

DESCRIPTION Small to tall tree. **Bark** rough over trunk and larger branches, thick, fibrous, grey-brown to red-brown. **Juvenile leaves** stalked, ovate to broad-lanceolate, to 15 x 8.5 cm. **Adult leaves** stalked, broad-lanceolate, to 16 x 4 cm, discolorous, dark green. **Buds** in axils of leaves, in 7s to 11s, stalked or stalk lacking, cylindrical to broadly spindle-shaped, to 0.9 x 0.6 cm; operculum conical to slightly beaked. **Flowers** white, December to February. **Fruit** usually lacking a stalk, cylindrical to barrel-shaped, to 1.2 x 0.9 cm.

DISTRIBUTION Coastal south-eastern Australia from near Newcastle on the mid coast of New South Wales to eastern Victoria in the Lakes Entrance area.

NOTES One of the red mahoganies, this species is widely planted for ornamental purposes. It is characterised by the thick rough bark and large dark green discolorous leaves.

Eucalyptus robusta

SWAMP MAHOGANY

DESCRIPTION Small to medium-sized tree. **Bark** rough over trunk and to the small branches, thick, fibrous, grey-brown to red-brown. **Juvenile leaves** stalked, ovate, to 19 x 8 cm. **Adult leaves** stalked, broad-lanceolate, to 17 x 4.5 cm, discolorous, dark green. **Buds** in axils of leaves, in > 7s, stalked, broadly spindle-shaped, to 2.4 x 0.8 cm; operculum beaked. **Flowers** white, May to July. **Fruit** stalked, cylindrical, to 1.8 x 1.1 cm, valves joined at their tips.

DISTRIBUTION Coastal eastern Australia, from near Nowra on the south coast of New South Wales northwards to north-west of Bundaberg in Queensland, including North Stradbroke, Moreton and Fraser Islands. Usually in swamps.

NOTES One of the red mahoganies, this species is characterised by the fruits which have the valves joined at their tips across the orifice. While naturally a swamp dweller, it has been used successfully as a street tree in southern Australia, including Perth.

Eucalyptus pellita

LARGE-FRUITED RED MAHOGANY

DESCRIPTION Small to tall tree. **Bark** rough over trunk and branches, thick, fibrous, grey-brown to red brown. **Juvenile leaves** stalked, ovate, to 21 x 8.5 cm. **Adult leaves** stalked, broad-lanceolate to slightly sickle-shaped, to 16 x 4 cm, discolorous, glossy green. **Buds** in axils of leaves, in > 7s, stalked or rarely stalk lacking, broadly spindle-shaped, to 2 x 1.2 cm; operculum conical, beaked or hemispherical. **Flowers** white, December to February. **Fruit** on stout stalks, funnel-shaped to hemispherical, to 1.4 x 1.7 cm, with exserted valves.

DISTRIBUTION Coastal and subcoastal ranges on the east side of Cape York Peninsula of northern Queensland; also New Guinea.

NOTES The northernmost red mahogany, it has been confused with the closely related southern species, *E. scias*. It is characterised by the rough bark and strongly discolorous leaves. A valuable commercial timber.

Eucalyptus longirostrata

GREY GUM

DESCRIPTION Medium-sized tree. **Bark** smooth but becoming somewhat granular with aging, shed in strips, mottled, cream, orange, coppery, grey and grey-brown. **Juvenile leaves** stalked, narrow-lanceolate to lanceolate, to 14.5 x 5 cm. **Adult leaves** stalked, lanceolate to slightly sickle-shaped, to 15 x 3 cm, discolorous, glossy, dark green. **Buds** in axils of leaves, in > 7s, stalked, to 2 x 0.8 cm; base funnel-shaped, operculum strongly beaked. **Flowers** white, December to April. **Fruit** stalked, hemispherical, to 1 x 1 cm, with exserted valves.

DISTRIBUTION South-eastern Queensland from north of Toowoomba to the Blackdown Tableland, including parts of the Great Dividing Range.

NOTES One of five grey gum species which are recognised by the mottled, +/– smooth bark and discolorous leaves.

Eucalyptus longifolia

WOOLLYBUTT

DESCRIPTION Medium-sized to tall tree. **Bark** rough over trunk and larger branches, thin, subfibrous and flaky, grey. **Juvenile leaves** stalked, ovate to broad-lanceolate, to 21 x 9 cm. **Adult leaves** stalked, lanceolate or falcate, to 24 x 2.5 cm, concolorous, green or greyish green. **Buds** pendulous, in 3s in axils of leaves, stalked, diamond-shaped, to 2.6 x 1.2 cm; operculum conical or beaked. **Flowers** white, October to November. **Fruit** stalked, cup-shaped, cylindrical or bell-shaped, pendulous, to 1.7 x 1.6 cm.

DISTRIBUTION South-eastern New South Wales from north of Newcastle to near the Victorian border. Coastal plains and low hills, often along drainage lines.

NOTES Easily recognised in south coastal New South Wales by the thin fibrous bark and the large, pendulous buds and fruits in 3s on long stalks.

46

Eucalyptus cosmophylla

CUP GUM

DESCRIPTION Mallee or small tree. **Bark** smooth, becoming minutely granular, or sometimes flaky, whitish to bluish grey. **Juvenile leaves** stalked, ovate to round, to 5 x 4 cm. **Adult leaves** stalked, lanceolate to broad-lanceolate, to 15 x 4 cm, concolorous, dull, light green. **Buds** in axils of leaves, in 3s, shortly stalked or lacking stalks, ovoid, to 1.8 x 1.3 cm; operculum conical, beaked or hemispherical. **Flowers** white, April to June. **Fruit** lacking a stalk, cup-shaped, to 2 x 1.8 cm.

DISTRIBUTION South Australia, southern Mt Lofty Ranges and Kangaroo Island, usually on poor soils.

NOTES Easily recognised as a small, often crooked tree, with relatively large buds and fruits in 3s.

47

Eucalyptus loxophleba

YORK GUM

DESCRIPTION Small to medium-sized tree. **Bark** rough over most or all of trunk, which may branch low, fibrous or flaky, grey-brown to dark brown. **Juvenile leaves** stalked, ovate, to 13 x 7 cm, bluish grey. **Adult leaves** stalked, lanceolate, to 13 x 2 cm, concolorous, glossy green. **Buds** formed at ends of branches, often maturing within the crown, in axils in 7s to 11s, stalked, club-shaped, to 0.9 x 0.4 cm; operculum hemispherical. **Flowers** white, September to January. **Fruit** stalked, funnel-shaped, to 0.9 x 0.6 cm.

DISTRIBUTION Western Australia, widespread in the south-west outside the wetter areas, from north of the Murchison to east of Jerramungup. Largely in cleared country in the western parts of the wheatbelt.

NOTES A very common tree, which may be confused in general appearance with other rough-barked species such as *E. longicornis*. Along roadsides the juvenile regrowth is always distinctive by the prominent, ovate, relatively large bluish grey leaves.

Eucalyptus kruseana

BOOK-LEAF MALLEE

DESCRIPTION Straggly mallee. **Bark** smooth, grey and pinkish grey. **Whole** mallee mature in juvenile leaf state. **Leaves** without stalks, round, crowded, to 2.2 x 2.3 cm, concolorous, dull grey. **Buds** in axils of leaves, in 7s, shortly stalked, club-shaped, to 0.9 x 0.4 cm; operculum conical or slightly beaked. **Flowers** yellow, June to September. **Fruit** shortly stalked, cup-shaped, to 0.9 x 0.8 cm.

DISTRIBUTION Western Australia, very restricted in distribution east and south-east of Kalgoorlie, at Cardunia Rock, Binyarinyinna Rock and a few other sites. Always on or near granite rocks.

NOTES Because of its small stature, crowded round grey leaves and yellow flowers, this species is a favourite ornamental in southern and central Australia.

49

Eucalyptus salubris

GIMLET

DESCRIPTION Small tree or mallee. **Bark** smooth, greenish grey to rich coppery; trunk with spirally twisted flutings. **Juvenile leaves** stalked, narrow-lanceolate, to 11 x 2 cm. **Adult leaves** stalked, narrow-lanceolate, to 11 x 1.5 cm, concolorous, glossy green. **Buds** in axils of leaves, stalked, broadly spindle-shaped, to 1.2 x 0.6 cm; operculum conical. **Flowers** white, December to March. **Fruit** stalked, cupular, to 0.6 x 0.7 cm.

DISTRIBUTION Western Australia, from the northern wheatbelt east to the western part of the Great Victoria Desert and south to the Fraser Range. Usually on heavy soils of plains and depressions.

NOTES A beautiful small tree with its spectacular coppery, fluted trunk and glossy green leaves.

Eucalyptus diptera

TWO-WINGED GIMLET

DESCRIPTION Mallee or small tree. **Bark** smooth, grey to rich coppery. **Juvenile leaves** stalked, ovate to broad-lanceolate, to 11 x 3.5 cm. **Adult leaves** stalked, narrow-lanceolate, to 11 x 1.5 cm, concolorous, glossy green. **Buds** in axils of leaves, in 3s, lacking stalks, crowded, squat, to 1.4 x 1.1 cm; operculum beaked. **Flowers** white, February to May. **Fruit** lacking stalks, hemispherical, two-ribbed, to 1.1 x 1.5 cm.

DISTRIBUTION Western Australia, subcoastal north and north-west of Esperance.

NOTES A beautiful small tree with fluted, coppery trunk, glossy green leaves and squat, crowded buds and fruits in 3s.

Eucalyptus annulata

DESCRIPTION Mallee or small tree. **Bark** smooth, green, greenish yellow or light coppery. **Juvenile leaves** stalked, lanceolate, to 5 x 4 cm. **Adult leaves** stalked, lanceolate, to 10 x 1.9 cm, concolorous, glossy, green. **Buds** in axils of leaves, in 7s, without stalks, elongated, to 3 x 0.7 cm; operculum long and widening at top. **Flowers** white, September to December. **Fruit** without stalks, almost globular but cut off at the top, to 0.8 x 1.2 cm, valves exserted.

DISTRIBUTION Subcoastal Western Australia from Dumbleyung in the west to north-east of Esperance.

NOTES A species related to the gimlets, *E annulata* is recognised by the thick, stubby buds and the dense clusters of fruits that remain held within the crown. The leaves are dark green and densely glandular, unlike the gimlets. Two closely related, more inland species have been described, *E. extensa* and *E. protensa*.

Eucalyptus grossa

THICK LEAF MALLEE

DESCRIPTION Straggly low mallee or small tree. **Bark** rough on stems and larger branches, light grey-brown. **Juvenile leaves** stalked, ovate, to 15 x 10 cm. **Adult leaves** stalked, ovate, to 15 x 4.5 cm, concolorous, thick, glossy green. **Buds** rigidly downturned in axils of leaves, in 7s, broadly spindle-shaped, tapering to stout stalks, to 3.3 x 1.2 cm; operculum conical. **Flowers** yellow-green, July to September. **Fruit** without stalks, cylindrical, to 2.2 x 1.4 cm.

DISTRIBUTION Southern Western Australia from south-east of Pingaring east towards Balladonia.

NOTES A species not resembling any other, *E grossa* is notable for its low form, thick glossy green leaves, and rigidly turned down buds and fruits.

Eucalyptus stricklandii

STRICKLAND'S GUM

DESCRIPTION Small tree. **Bark** rough on lower part of trunk, dark grey to black, smooth white, grey or pinkish above. **Juvenile leaves** stalked, elliptical to ovate, to 8 x 5 cm. **Adult leaves** stalked, lanceolate, to 17 x 2.8 cm, concolorous, glossy green. **Buds** in axils of leaves, in 7s, without stalks or on short stout stalks, cylindrical and flared around the middle, to 2.4 x 1.2 cm; operculum +/– bell-shaped. **Flowers** yellow, November to March. **Fruit** without stalks, bell-shaped (inverted), to 1.7 x 1.5 cm.

DISTRIBUTION Western Australia, from Diemals north of Coolgardie south and south-east towards Norseman and east towards Zanthus, usually on stony ground.

NOTES A popular ornamental in southern Australia, with its yellow flowers.

Eucalyptus dielsii

CAP-FRUITED MALLEE

DESCRIPTION Mallee or small tree. **Bark** smooth, grey, greenish grey, whitish grey or coppery. **Juvenile leaves** stalked, ovate to heart-shaped, to 8 x 6 cm. **Adult leaves** stalked, lanceolate, to 12 x 2.2 cm, concolorous, glossy, olive-green to green. **Buds** pendulous, in axils of leaves, in 7s, ovoid, to 1.7 x 0.9 cm; operculum conical to slightly beaked. **Flowers** pale green to yellow-green, December to January. **Fruit** stalked, hanging down, cup-shaped at the base but bell-shaped overall with a flared rim, to 1.2 x 1.2 cm.

DISTRIBUTION Subcoastal southern Western Australia from north-west of Ravensthorpe to north-east of Esperance, usually on low clayey ground.

NOTES An attractive ornamental in southern Australia, with its smooth, often coppery bark and large clusters of beautiful greenish flowers.

55

Eucalyptus erythronema

RED-FLOWERED MALLEE

DESCRIPTION Mallee. **Bark** smooth, powdery, varying from white to purplish or red. **Juvenile leaves** stalked, lanceolate, to 10 x 2 cm. **Adult leaves** stalked, narrow-lanceolate, to 9 x 1.4 cm, concolorous, glossy, olive-green to green. **Buds** pendulous, in axils of leaves, in 3s or 7s, ovoid or diamond-shaped, to 2.2 x 0.9 cm; operculum conical or beaked. **Flowers** red, pink or creamy white, October to January. **Fruit** stalked, hanging down, funnel-shaped, often with flared rim, to 1.4 x 1.3 cm.

DISTRIBUTION In the wheatbelt of Western Australia extending east to Southern Cross.

NOTES A beautiful ornamental mallee with its spectacular coloured bark in season and red flowers.

Eucalyptus platypus

MOORT

DESCRIPTION Small tree. **Bark** smooth, grey to red-brown. **Juvenile leaves** stalked, ovate, to 7 x 4 cm. **Adult leaves** stalked, ovate to round, sometimes notched at tip, to 7.5 x 3.5 cm, concolorous, glossy, olive-green to green. **Buds** in axils of leaves, in 7s, without stalks or on short, stout stalks, elongated, to 3 x 0.9 cm; operculum horn-shaped. **Flowers** white, September to January. **Fruit** without stalks or on short, stout stalks, funnel-shaped, to 1.8 x 1.4 cm.

DISTRIBUTION Coastal and subcoastal southern Western Australia between Albany and Esperance. Usually growing in pure stands of even height on heavy ground.

NOTES A species easy to identify with its characteristic form and site and, in particular, the roundish leaves and long horn-shaped operculum of the bud.

Eucalyptus spathulata

SWAMP MALLET

DESCRIPTION Mallee or small tree. **Bark** smooth, satiny, grey to salmon-pink. **Juvenile leaves** stalked, linear to narrow-lanceolate, to 5 x 0.5 cm. **Adult leaves** stalked, linear, to 10 x 0.5 cm, concolorous, glossy, olive-green, crowded with oil glands. **Buds** in axils of leaves, in 7s, stalked, cylindrical, to 1.3 x 0.4 cm; operculum cylindrical, rounded at top. **Flowers** white, December to March. **Fruit** stalked, hemispherical to funnel-shaped, to 0.7 x 0.7 cm.

DISTRIBUTION South-west Western Australia from Tammin east of Perth in the north south-east towards Ongerup. Usually on heavy red-soiled flats.

NOTES A popular street tree in southern Australia, with its attractive fine foliage and smooth stems.

Eucalyptus steedmanii

STEEDMAN'S MALLET

DESCRIPTION Small tree or mallet. **Bark** smooth, grey to bright coppery. **Juvenile leaves** stalked, ovate to broad-lanceolate, to 9 x 3 cm. **Adult leaves** stalked, elliptical to lanceolate, to 8 x 1.5 cm, concolorous, glossy, olive-green to green, crowded with oil glands. **Buds** pendulous, in axils of leaves, in 3s, on long stalks, elongated diamond-shaped, square in cross-section, to 3.3 x 1.3 cm; operculum pyramidal. **Flowers** white, January to March. **Fruit** on long stalks, square in cross-section and longitudinally winged, to 2.2 x 1.7 cm.

DISTRIBUTION Of very restricted distribution in southern Western Australia, occurring in only a few stands east of Hyden.

NOTES This species was discovered in 1928, described in 1933, then 'lost' for many years, its locality being only poorly documented. In the meantime from original seed collections, it became well known in southern Australia as an ornamental tree. It was found again in the 1980s and its known distribution considerably enlarged, particularly to the north of the site where it was discovered.

Eucalyptus eremophila

SAND MALLEE

DESCRIPTION Mallee or small tree. **Bark** smooth, grey-white, pink or brilliant coppery. **Juvenile leaves** stalked, ovate, to 11 x 4 cm. **Adult leaves** stalked, narrow-lanceolate to lanceolate, to 12 x 1.5 cm, concolorous, glossy, olive-green to green, crowded with oil glands. **Buds** pendulous, in axils, in 7s or 9s, stalked, elongated, to 4 x 0.7 cm; operculum horn-shaped. **Flowers** creamy, pale yellow or rarely pale pink, August to December. **Fruit** stalked, cylindrical to cup-shaped, to 1.5 x 1.2 cm.

DISTRIBUTION South-west Western Australia from east of Dongara to the south coast and east to Zanthus on the Trans Australian Railway. Usually on sand.

NOTES An attractive small mallee or tree with its colourful bark, glossy leaves and pendulous buds and fruits.

60

Eucalyptus occidentalis

SWAMP YATE

DESCRIPTION Small to medium-sized tree. **Bark** rough on lower trunk, dark grey-brown to black, upper trunk and branches smooth, white. **Juvenile leaves** stalked, ovate, to 14 x 7 cm. **Adult leaves** stalked, lanceolate, to 16 x 3.3 cm, concolorous, glossy green. **Buds** pendulous, in axils of leaves, in 7s, rarely 9s, stalked, elongated, to 3.3 x 0.7 cm; operculum horn-shaped. **Flowers** white, November to May. **Fruit** stalked, bell-shaped, to 1.5 x 1.2 cm, valves exserted.

DISTRIBUTION Southern coastal and subcoastal Western Australia, in the wheatbelt east to the Mt Ragged area. Usually around freshwater swamps.

NOTES A tree of the yate group, it is easy to recognise because of its watery habitat, rough basal bark, green leaves and pendulous buds and fruit.

Eucalyptus astringens

BROWN MALLET

DESCRIPTION Small to medium-sized tree or mallet. **Bark** mostly smooth, with loose, curly flakes, grey, brown or coppery. **Juvenile leaves** stalked, ovate, to 10 x 6 cm. **Adult leaves** stalked, narrow-lanceolate, to 11 x 2.3 cm, concolorous, glossy green. **Buds** in axils of leaves, in 7s, finally pendulous, stalked, elongated, to 2 x 0.6 cm; operculum cylindrical with rounded tip. **Flowers** white, September to December. **Fruit** stalked, hanging down, bell-shaped, to 1.2 x 1 cm.

DISTRIBUTION South-western Western Australia, from west of Brookton in the southern wheatbelt south to the southern coast. Usually on hills and particularly ironstone breakaways.

NOTES One of the mallets that characteristically occupy breakaways. It is distinguished from another mallet of similar habit, *E. gardneri*, which is strikingly blue-grey leaved and has long pointed buds.

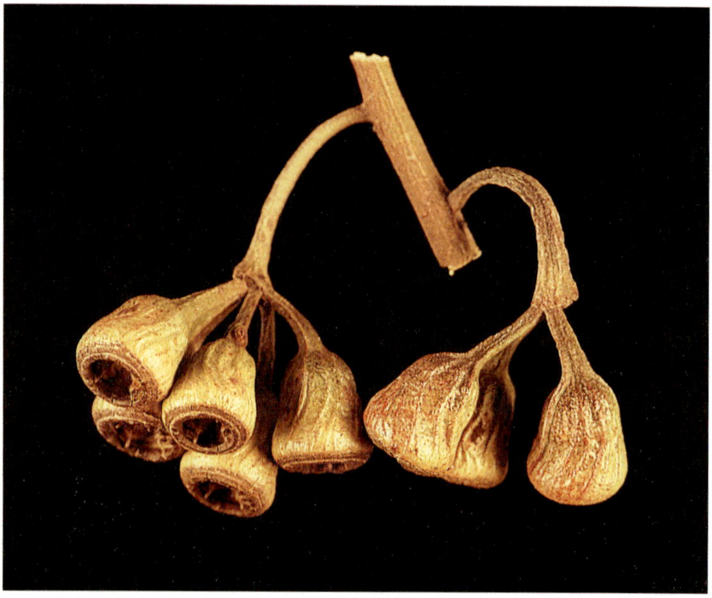

62

Eucalyptus cornuta

YATE

DESCRIPTION Small to tall tree or rarely mallee. **Bark** rough on lower half of trunk, hard, compacted and furrowed, dark grey to grey-black. **Juvenile leaves** stalked, round, to 5 x 5 cm. **Adult leaves** stalked, elliptical, to 14 x 3.5 cm, concolorous, glossy, dark green to grey-green. **Buds** in axils of leaves, in more than 7s, without stalks and tightly clustered at the base, elongated, to 4.2 x 0.8 cm; operculum horn-shaped. **Flowers** yellowish, July to November. **Fruit** without stalks, crowded, cup-shaped, to 1.3 x 1.3 cm; valves joined across the opening.

DISTRIBUTION Far south Western Australia, coastal, from Hamelin Bay east to Orleans Bay and islands of the Recherche Archipelago.

NOTES A long-established ornamental tree in southern Australia, the populations of the tallest trees in their natural habitats have been largely cleared because of their excellent timber.

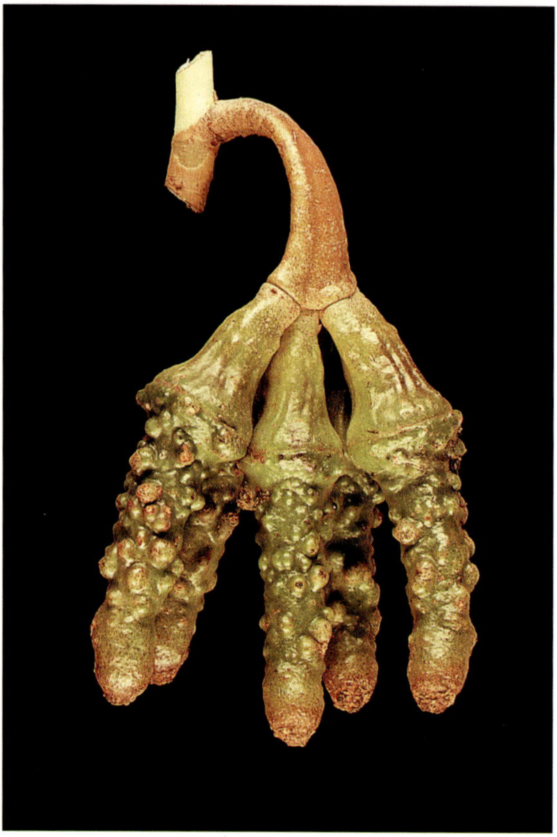

Eucalyptus megacornuta

WARTED YATE

DESCRIPTION Small tree or mallet. **Bark** smooth, grey, brownish grey or orange. **Juvenile leaves** stalked, petiolate, lanceolate, to 8 x 6 cm. **Adult leaves** stalked, lanceolate, to 11 x 2.2 cm, concolorous, glossy green. **Buds** pendulous, in axils of leaves, in 7s, rarely 3s, without stalks or base of bud narrowing to short stout stalk, elongated, to 6.8 x 2.3 cm; operculum horn-shaped, warty. **Flowers** yellow-green, July to December. **Fruit** without stalks or stalks short, stout, bell-shaped, to 3.5 x 3.2 cm; rim very thick, with exserted valves.

DISTRIBUTION Southern Western Australia, known from only a few sites in the Ravensthorpe Range. Always on stony slopes.

NOTES A popular ornamental small tree in southern Australia, noted for its grossly warted buds and beautiful large yellow-green flower masses.

Eucalyptus conferruminata

BALD ISLAND MARLOCK

DESCRIPTION Small tree or mallee. **Bark** smooth, grey and whitish grey. **Juvenile leaves** stalked, ovate to orbicular, to 10 x 4 cm. **Adult leaves** stalked, elliptical, to 9 x 2.5 cm, concolorous, glossy, light green. **Buds** in axils of leaves, to > 20-flowered, hanging down, all fused at their bases with long horn-shaped opercula to 6 cm long. **Flowers** yellow-green, August to November. **Fruit** fused in a tight ball hanging conspicuously inside crown, to 6 x 8 cm; rim thick with exserted valves.

DISTRIBUTION Southern Western Australia, from Two Peoples Bay east to beyond Esperance, including islands. Always coastal, often on massive granite rocks.

NOTES A relatively recently described species, *E. conferruminata* was incorrectly known until 1981 as *E. lehmannii,* which is a smaller, finer-leaved, more inland mallee. It has been widely planted in southern Australia, particularly in coastal areas, where it forms dense crowns and is effective as a windbreak when planted in rows.

Eucalyptus wandoo

WANDOO

DESCRIPTION Small to medium-sized tree. **Bark** on saplings rough and fibrous; on trees, smooth white, grey, yellow, or pale orange. **Juvenile leaves** stalked, heart-shaped to broad-lanceolate, to 13 x 10 cm. **Adult leaves** stalked, lanceolate, to 15 x 2.4 cm, concolorous, dull, light green to grey-green. **Buds** in axils of leaves, in > 7s, stalked, spindle-shaped, to 1.9 x 0.4 cm; operculum narrowly conical. **Flowers** white, December to May. **Fruit** stalked, cup-shaped to cylindrical, to 1 x 0.6 cm.

DISTRIBUTION South-western Western Australia, in the Darling Range and adjacent foothills from Gin Gin and Bindi Bindi in the north, south towards Donnybrook and south-east to the Stirling Range and Pallinup River.

NOTES A common, mostly white-barked tree largely in low hilly country. There is a related form, *E. capillosa,* in the wheatbelt that differs most notably in having hairy juvenile leaves and more colourful bark.

Eucalyptus xanthonema

SMALL-LEAVED MALLEE

DESCRIPTION Small mallee. **Bark** smooth, dark grey or pinkish grey. **Juvenile leaves** stalked, broad-lanceolate to ovate, to 6 x 2 cm. **Adult leaves** stalked, narrow-lanceolate to linear, thin, to 9 x 0.6 cm, concolorous, dull to slightly glossy, green. **Buds** in axils of leaves, in 7s to 11s, stalked, spindle-shaped, to 1.7 x 0.3 cm; operculum narrowly conical, with tip usually bent. **Flowers** white, September to February. **Fruit** stalked, cup-shaped, to 0.5 x 0.5 cm.

DISTRIBUTION Southern wheatbelt of Western Australia, from south-east of Williams to the middle of Fitzgerald River National Park.

NOTES This species is always a low mallee with a dense crown of small thin leaves.

Eucalyptus gardneri

BLUE MALLET

DESCRIPTION Small to medium-sized tree or mallet. **Bark** smooth, grey, light coppery or silvery white, with characteristic curling flakes of partly shed bark. **Juvenile leaves** stalked, ovate, to 10 x 6 cm. **Adult leaves** stalked, lanceolate, to 9 x 2.3 cm, concolorous, dull, blue-green to purplish. **Buds** in axils of leaves, in > 7s, stalked, spindle-shaped, to 2.6 x 0.5 cm; operculum long, horn-shaped, hooked at the top. **Flowers** pale yellow, March to November. **Fruit** stalked, funnel-shaped or barrel-shaped, to 1.1 x 0.7 cm.

DISTRIBUTION West central wheatbelt of southern Western Australia, from Brookton east to Narambeen and south to Harrismith; also a small occurrence east of Cadoux in the northern wheatbelt.

NOTES An attractive small tree with purplish foliage in season and yellowish flowers.

68

Eucalyptus desmondensis

DESMOND MALLEE

DESCRIPTION Spindly mallee. **Bark** smooth, grey to pale orange. **Juvenile leaves** stalked, ovate to heart-shaped, to 8 x 6 cm. **Adult leaves** stalked, lanceolate to broad-lanceolate, to 10 x 3.5 cm, concolorous, slightly glossy, green. **Buds** in axils of leaves, in > 7s, without stalks, broadly spindle-shaped, to 1.1 x 0.6 cm; operculum conical. **Flowers** pale yellow, February to June or irregular. **Fruit** without stalks, cup-shaped to cylindrical, to 1.1 x 0.9 cm.

DISTRIBUTION Ravensthorpe Range of southern Western Australia, particularly on granite sands.

NOTES A small mallee recognised by its poor spindly form, often drooping crowns, stems dark red and shiny with white waxy overlay, and yellowish flowers.

Eucalyptus dundasii

DUNDAS BLACKBUTT

DESCRIPTION Small to medium-sized tree. **Bark** rough on lower trunk, tessellated, dark brown to brown-black, smooth whitish, grey or coppery above. **Juvenile leaves** stalked, ovate, to 15 x 5 cm, glossy green. **Adult leaves** stalked, lanceolate to narrow-lanceolate, to 12 x 1.6 cm, concolorous, glossy green. **Buds** in axils of leaves, in 7s, without stalks or shortly stalked, cylindrical to pear-shaped, to 1 x 0.5 cm; operculum beaked. **Flowers** white, February to May. **Fruit** without stalks or shortly stalked, cylindrical, to 1 x 0.5 cm.

DISTRIBUTION Southern Western Australia, with the main occurrence from north-west of Norseman to the Fraser Range and south to Kumarl, with a smaller occurrence north of Coolgardie between Mulline and Callion.

NOTES A tree planted in drier areas of southern Australia, this species is recognised by the characteristic dark, tessellated rough basal bark and the bright glossy green leaves at all stages.

Eucalyptus doratoxylon

SPEARWOOD

DESCRIPTION Mallee. **Bark** smooth, pinkish, grey-brown or white and powdery, sometimes with a rough stocking on large stems. **Juvenile leaves** without stalks, opposite, lanceolate, to 6 x 1 cm, green. **Adult leaves** shortly stalked, opposite, lanceolate, to 8 x 1.1 cm, glossy, dark green. **Buds** pendulous, in axils of leaves, in 7s, stalked; base cup-shaped, operculum beaked, to 0.9 x 0.4 cm. **Flowers** white, August to November. **Fruit** stalked, hanging down, barrel-shaped, to 0.8 x 0.7 cm.

DISTRIBUTION Southern Western Australia in coastal and subcoastal areas from east of Albany and the Stirling Range east to Cape Arid National Park.

NOTES A delicate mallee unique among southern species for having the leaves of the mature crown opposite. The bark is intensely white in season.

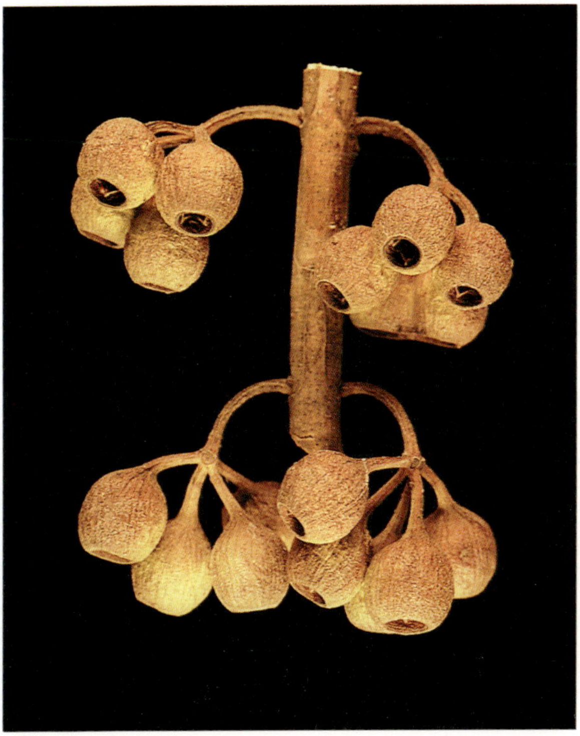

Eucalyptus falcata

SILVER MALLEE

DISTRIBUTION South-west Western Australia outside the wetter areas, from north-east of Dongara to Esperance on the south coast.

NOTES A small mallee with attractive silvery crown and delicate, pendulous, cream-coloured buds and fruits.

DESCRIPTION Mallee. **Bark** smooth, grey, pink, red-brown or silver. **Juvenile leaves** stalked, ovate, to 6 x 3 cm. **Adult leaves** stalked, lanceolate or sickle-shaped, to 14 x 2 cm, concolorous, glossy, green to dark green, but whole crown often silvery. **Buds** pendulous, in axils of leaves, double-conic in 7s to 11s, stalked, often ribbed, to 1.8 x 0.8 cm; operculum long conical. **Flowers** white, September to March. **Fruit** stalked, hanging down, almost spherical, smooth or ribbed, to 0.8 x 1.1 cm.

Eucalyptus mannensis

MANN RANGE MALLEE

DESCRIPTION Mallee. **Bark** rough over most of stems, loose, flaky, grey to yellow-brown, smooth grey, yellowish or red-brown above. **Juvenile leaves** stalked, elliptical to lanceolate, to 9 x 3 cm. **Adult leaves** stalked, lanceolate, to 14.5 x 1.7 cm, concolorous, glossy green, densely glandular. **Buds** in axils of leaves, in 7s, shortly stalked, ovoid, to 0.9 x 0.5 cm; operculum conical. **Flowers** white, April to October. **Fruit** shortly stalked, hemispherical, to 0.7 x 0.9 cm; rim thick; valves exserted.

DISTRIBUTION Scattered but widespread in central and eastern Western Australia, south-western Central Australia and north-western South Australia. Usually in sandy country.

NOTES A mallee conspicuous in desert areas because of its bright glossy green leaves, which yield high quantities of eucalyptus oil.

Eucalyptus cneorifolia

KANGAROO ISLAND NARROW-LEAVED MALLEE

DESCRIPTION Mallee or rarely small tree. **Bark** rough over most of stems, grey-brown. **Juvenile leaves** shortly stalked, sub-opposite, narrow-lanceolate to lanceolate, to 5 x 1 cm. **Adult leaves** shortly stalked, sub-opposite to alternate, linear to narrow-lanceolate, held erect, to 12 x 1 cm, concolorous, glossy green. **Buds** in axils of leaves, in 7s to 11s, without stalks or shortly stalked, broadly spindle-shaped, to 1 x 0.5 cm; operculum conical. **Flowers** white, March to May. **Fruit** without stalks, crowded, hemispherical, to 0.7 x 0.8 cm.

DISTRIBUTION Eastern end of Kangaroo Island and near Waitpinga on Fleurieu Peninsula of South Australia.

NOTES A mallee often occurring in dense, pure stands and recognised by the rough bark and narrow, erect leaves. It is harvested for eucalyptus oil extraction.

Eucalyptus cooperiana

ESPERANCE MALLEE

DESCRIPTION Mallee. **Bark** smooth, red-brown to powdery white. **Juvenile leaves** without stalks and opposite for many pairs, then stalked, alternate, ovate to round, to 10 x 6 cm. **Adult leaves** stalked, lanceolate, to 11 x 2.4 cm, concolorous, glossy green. **Buds** erect or pendulous, in axils of leaves, in 7s to 15s, stalked, with urn-shaped base and short rounded operculum, to 0.9 x 0.5 cm. **Flowers** white, January to March. **Fruit** stalked, hanging down, urn-shaped, to 0.9 x 0.9 cm.

DISTRIBUTION Restricted to the sub-coastal sandplain between Esperance and Israelite Bay, Western Australia.

NOTES An attractive mallee with its usually stark white stems, glossy green leaves and pendulous buds and fruit.

Eucalyptus angustissima

FINE-LEAVED MALLEE

DESCRIPTION Mallee. **Bark** smooth, grey, whitish grey or light grey-brown. **Juvenile leaves** without stalks or shortly stalked, sub-opposite to alternate, linear, to 7 x 0.6 cm. **Adult leaves** shortly stalked, linear, held erect, to 13 x 0.4 cm, concolorous, glossy green. **Buds** in axils of leaves, in 7s, stalked, ovoid, to 0.6 x 0.5 cm; operculum conical, narrower than base. **Flowers** white, August to January. **Fruit** shortly stalked, hemispherical, to 0.7 x 0.6 cm.

DISTRIBUTION Far southern Western Australia from Lake Chinacup east to Israelite Bay.

NOTES The mallee with narrowest leaves of all eucalypt species.

Eucalyptus salmonophloia

SALMON GUM

DESCRIPTION Small to medium-sized tree. **Bark** smooth, white, silver grey or in season, salmon pink to coppery. **Juvenile leaves** stalked, alternate, ovate to lanceolate, to 9 x 3 cm. **Adult leaves** stalked, narrow-lanceolate to lanceolate, to 12 x 1.5 cm, concolorous, glossy green. **Buds** in axils of leaves, in 7s or more, with fine stalks, ovoid to globular, to 0.7 x 0.4 cm; operculum hemispherical. **Flowers** white, September to December. **Fruit** stalked, hemispherical, to 0.5 x 0.5 cm; valves exserted.

DISTRIBUTION Widespread throughout the wheatbelts of Western Australia from Morowa in the north-west, east to Pinjin Station, south to Ongerup and south-east of the Fraser Range. Always on plains.

NOTES One of the most beautiful trees outside of the high-rainfall country of the south-west, with its spectacular coloured bark in season and bright green glossy leaves.

Eucalyptus oleosa

RED MALLEE

DESCRIPTION Mallee. **Bark** rough for the lower 1–2 m of the stems, grey-brown. **Juvenile leaves** without stalks, scattered around the stems in spirals, linear, to 4 x 0.5 cm. **Adult leaves** stalked, narrow-lanceolate, to 10 x 1 cm, concolorous, glossy green. **Buds** in axils of leaves, in 7s to 13s, stalked, ovoid, with the operculum characteristically narrow giving an egg-in-eggcup appearance, to 0.8 x 0.3 cm. **Flowers** white, December to June. **Fruit** stalked, slightly urn-shaped, to 0.6 x 0.6 cm; slender style parts remain exserted for some time, often joined at the top.

DISTRIBUTION Widespread in the mallee scrubs of southern Australia, from the goldfields of Western Australia east to central New South Wales.

NOTES A species name traditionally embracing many forms but correctly applied to rough-barked mallees, with glossy green crowns, no glands in the pith of the branchlets, egg-in-eggcup shaped buds, and fruits with the fragile exserted style tips.

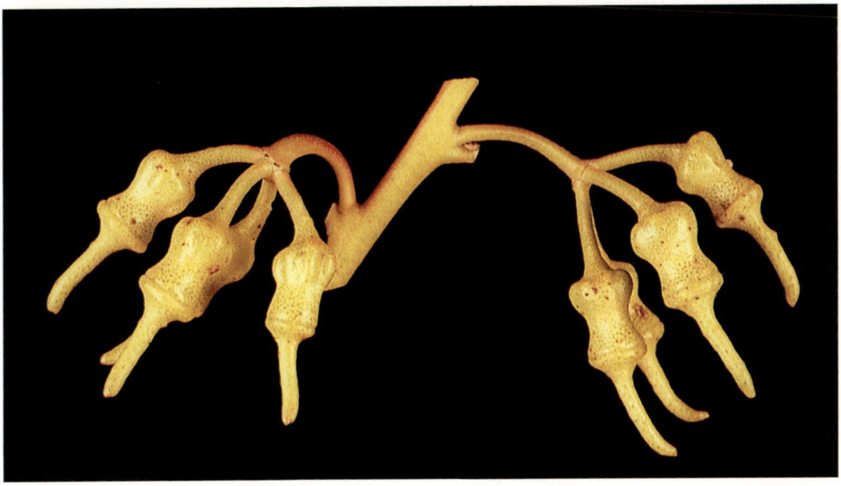

Eucalyptus flocktoniae

FLOCKTON'S MALLEE

DESCRIPTION Small tree, mallet or mallee. **Bark** smooth, grey, creamy, silvery or coppery. **Juvenile leaves** without stalks, opposite with leaf edges continuing as narrow wings down part of the stem, to 12 x 3.5 cm. **Adult leaves** stalked, lanceolate, to 14 x 2.2 cm, concolorous, glossy green. **Buds** usually pendulous in axils of leaves, in 7s to 11s, stalked; base urn-shaped with prominent 'waist', operculum strongly beaked, to 1.9 x 0.6 cm. **Flowers** white, October to April. **Fruit** stalked, hanging down, very variable in shape from urn-shaped to almost hemispherical, to 1 x 0.8 cm.

DISTRIBUTION Widespread in southern Western Australia, from near Three Springs east towards Caiguna, and also on Eyre Peninsula in South Australia.

NOTES A very variable species particularly in the shape and sculpturing of the buds and fruits. Crown is always glossy green.

Eucalyptus oxymitra

SHARP-CAPPED MALLEE

DESCRIPTION Mallee. **Bark** smooth, grey, whitish grey or cream, usually with partly shed ribbons of dead bark at base. **Juvenile leaves** stalked, ovate to broad-lanceolate, to 11 x 4.5 cm. **Adult leaves** stalked, broad-lanceolate to lanceolate, to 11 x 3 cm, concolorous, dull, blue-grey to glaucous, firm texture. **Buds** in axils of leaves, in 7s, stalked, with hemispherical base and strongly beaked operculum, to 1.8 x 1 cm. **Flowers** white, December to January. **Fruit** stalked, with hemispherical base, ascending disc and strongly exserted valves, to 1.3 x 2 cm.

DISTRIBUTION East central Western Australia, south-western Central Australia and north-western South Australia to as far as the Everard Range. Confined to desert areas of sandy spinifex plains or stony hills, e.g. Mt Connor.

NOTES It is distinguished from other mallees in the area by the dull, thickish, greyish leaves, the glaucous buds and fruits, and the sharp, pointed valves of the fruit.

Eucalyptus rameliana

GILES'S MALLEE

DESCRIPTION Low sprawling mallee. **Bark** smooth, grey and pinkish grey, or sometimes rough at base. **Juvenile leaves** stalked, broadly ovate, to 7 x 5.5 cm. **Adult leaves** stalked, ovate, to 11 x 4 cm, concolorous, dull, blue-green to grey-green, firm texture. **Buds** pendulous in axils of leaves, single or rarely in 3s, stalked; base funnel-shaped and operculum conical or beaked, to 4 x 2.5 cm. **Flowers** yellow, rarely pink to red, May to June. **Fruit** hanging down, stalked, hemispherical with broad ascending disc enclosing 4 exserted valves, to 2.5 x 3.3 cm.

DISTRIBUTION Southern part of the Great Sandy Desert in central Western Australia, south-east of Newman. On red desert dunes.

NOTES This species, found in 1876, was rediscovered only in 1991. Believed to be extinct for many years, it has now been found to occur over a large area of remote country. It is one of the few species that has a single flower in its inflorescence. Its performance in cultivation is unknown but eagerly awaited.

Eucalyptus pyriformis

DOWERIN ROSE

DESCRIPTION Mallee. **Bark** smooth, grey or salmon pink, often with ribbons at base. **Juvenile leaves** stalked, alternate, ovate, to 9 x 5 cm. **Adult leaves** stalked, broad-lanceolate, to 9.5 x 3.2 cm, concolorous, dull, grey to grey-green, firm texture. **Buds** pendulous in axils of leaves, in 3s, on long thick stalks, club-shaped, ribbed, to 6 x 3.3 cm; operculum hemispherical with a terminal point. **Flowers** red or creamy white, July to October. **Fruit** hanging down on long stalks, funnel-shaped, ribbed, to 4 x 5.5 cm; rim thick, usually with strongly ascending concave disc.

DISTRIBUTION The western part of the northern wheatbelt of Western Australia, from north-west of Geraldton in the north to about Dowerin in the south. Usually on white sandplain.

NOTES Usually a straggly small mallee with large pendulous buds and fruits and spectacular coloured flowers.

Eucalyptus youngiana

YARLDARLBA

DESCRIPTION Mallee. **Bark** rough, loose, flaky over part or most of stems, grey-brown to yellow-brown, rarely wholly smooth. **Juvenile leaves** stalked, alternate, elliptical to broadly ovate, to 12 x 8 cm. **Adult leaves** stalked, broad-lanceolate, to 17 x 3.5 cm, concolorous, dull, light green to blue green. **Buds** in axils of leaves, in 3s, on short, stout stalks, subspherical with pointed operculum, ribbed, to 5 x 5 cm. **Flowers** red or yellow, June to October. **Fruit** on short, stout stalks, hemispherical, ribbed, to 4.5 x 7 cm; rim thick with strongly ascending, concave disc.

DISTRIBUTION Widespread in southern Australia in the Great Victoria Desert and fringes, from north of Kalgoorlie east across the north of the Nullarbor Plain in South Australia.

NOTES A beautiful ornamental for arid areas, with its large buds and fruits and spectacular coloured flowers.

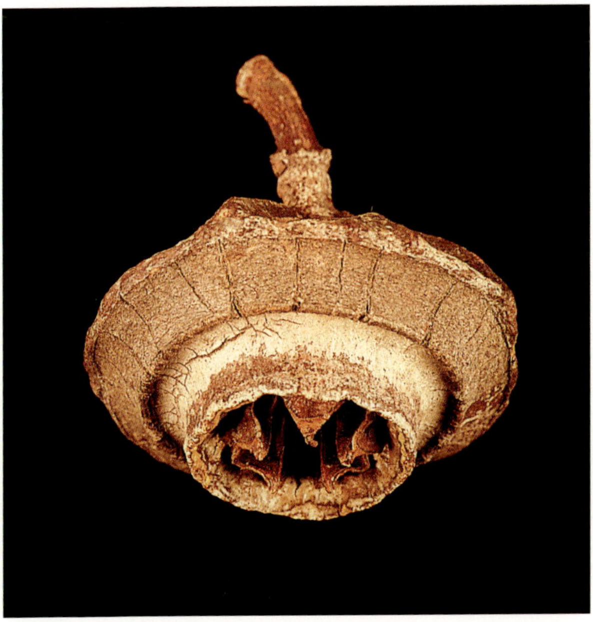

Eucalyptus kingsmillii

KINGSMILL'S MALLEE

DESCRIPTION Mallee. **Bark** usually rough, loose, over most of stems, light grey to grey-brown, rarely completely smooth. **Juvenile leaves** stalked, ovate to broad-lanceolate, to 13 x 5 cm. **Adult leaves** stalked, broad-lanceolate, to 15 x 3 cm, concolorous, dull, blue-green to blue-grey. **Buds** pendulous in axils of leaves, in 3s, stalked; base cup-shaped or hemispherical, operculum strongly beaked, red or yellow, ribbed, to 5 x 3.5 cm. **Flowers** creamy white to yellow, April to September. **Fruit** stalked, cup-shaped to hemispherical, ribbed, to 3 x 4 cm; rim very thick with broad, ascending, concave disc.

DISTRIBUTION Western Australia, very scattered but widespread in central western arid areas extending eastwards to the Gibson Desert, also on hill and mountain tops in the Pilbara.

NOTES A beautiful, usually small mallee of desert sandy areas but also stony sites in the Pilbara, notable for its large, colourful, pendulous buds and fruits.

Eucalyptus pachyphylla

RED BUD MALLEE

DESCRIPTION Mallee. **Bark** rough, loose ,ribbony on lower part of stems, or wholly smooth, white, pink or red-brown. **Juvenile leaves** stalked, ovate to broad-lanceolate, to 16 x 5 cm. **Adult leaves** stalked, broad-lanceolate, to 16 x 4 cm, concolorous, dull, blue-grey. **Buds** in axils of leaves, in 3s, stalked; base shallowly hemispherical, operculum strongly beaked, smooth or ribbed, red or cream, to 4.5 x 2.6 cm. **Flowers** pale yellow, April to July. **Fruit** stalked, shallowly hemispherical, ribbed, to 2 x 3.3 cm; rim very thick with broad, ascending, concave disc.

DISTRIBUTION Scattered in the Gibson Desert of central eastern Western Australia east and north-east across the Northern Territory to far western Queensland.

NOTES A beautiful small mallee with large colourful buds and pale yellow flowers. It is closely related to *E. kingsmillii,* from which it differs in the shorter inflorescence stalks and bud stalks and erect buds and fruit.

85

Eucalyptus crucis

SOUTHERN CROSS MALLEE

DESCRIPTION Sprawling mallee or small tree. **Bark** partly shed and held in long, narrow curled strips with underbark smooth, rich coppery red (minniritchi). **Juvenile leaves** without stalks, opposite for many pairs, ovate to round, to 4 x 5 cm, concolorous, white, waxy. **Mature crown** composed of juvenile leaves only. **Buds** in axils of leaves, in 7s, stalked, ovoid, white, waxy, to 1.1 x 0.8 cm; operculum obtusely conical. **Flowers** white, December to March. **Fruit** stalked, hemispherical, white, waxy, to 0.9 x 1.6 cm; rim very thick with broad, ascending disc.

DISTRIBUTION Confined to a small area of south-western Western Australia between Perth and Coolgardie e.g. Sanford Rock. Always on or around granite rocks.

NOTES A popular ornamental with its beautiful rich-coloured sculptured bark and complete crown of glaucous, roundish leaves.

Eucalyptus orbifolia

ROUND-LEAVED MALLEE

DESCRIPTION Mallee. **Bark** minniritchi (see right). **Juvenile leaves** stalked, round, to 5 x 6 cm, greyish. **Adult leaves** stalked, somewhat oval but notched at the end, to 5.5 x 3.7 cm, concolorous, dull, grey-green to white waxy. **Buds** in axils of leaves, in 7s, stalked, almost globular but with rounded or pointed operculum, white waxy, to 1 x 0.8 cm. **Flowers** yellow, July to November. **Fruit** stalked, hemispherical, white waxy, to 1 x 1.4 cm; rim very thick with broad, ascending disc.

DISTRIBUTION The northern goldfields of Western Australia and McDonnell Range and other high country in south-western Central Australia and far north-western South Australia.

NOTES A fine ornamental with its low stature, coloured sculptured bark and broad whitish leaves.

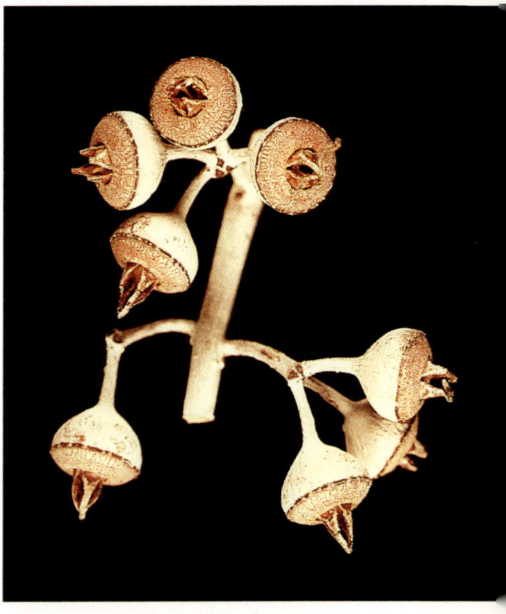

Eucalyptus caesia

CAESIA

DESCRIPTION Mallee with two forms, one with small leaves, buds and fruits (subspecies *caesia*), the other with large leaves, buds and fruits (subspecies *magna*). **Bark** partly shed and held in long, curled narrow strips with underbark smooth, rich, coppery red (minniritchi). **Juvenile leaves** stalked, alternate, heart-shaped, to 8 x 6 cm (*caesia*), to 10 x 8 cm (*magna*), glossy green. **Adult leaves** stalked, lanceolate, to 13.5 x 2.6 cm (*caesia*), to 24 x 5 cm (*magna*), concolorous, dull, light green to blue-green. **Buds** pendulous in axils of leaves, in 3s, stalked, club-shaped, to 2.7 x 1.1 cm (*caesia*), to 4 x 2 cm (*magna*); operculum conical or beaked. **Flowers** pink (*caesia*) or red (*magna*), May to August. **Fruit** stalked, hanging down, cup-shaped to urn-shaped, to 2.6 x 2.2 cm (*caesia*), to 3.7 x 4 cm (*magna*).

DISTRIBUTION Western Australia, confined to granite rocks in the wheatbelt, from Boyagin Rock to The Humps north-east of Hyden.

NOTES Planted widely in southern Australia, these mallees are among the most beautiful ornamentals, particularly the larger form (*magna*) with its large, greyish leaves, and large pendulous buds, fruits and coloured flowers. It is notable for its winter flowering.

Eucalyptus leptophylla

NARROW-LEAVED RED MALLEE

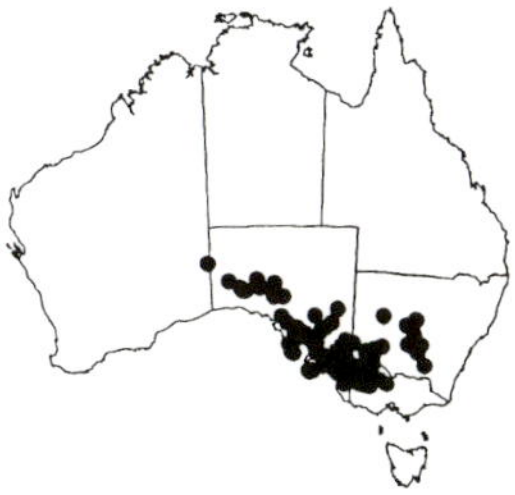

DESCRIPTION Mallee. **Bark** smooth, grey. **Juvenile leaves** without stalks, opposite for many pairs, elliptical to ovate, to 2.5 x 1.6 cm, greyish. **Adult leaves** stalked, narrow-lanceolate, to 8 x 0.8 cm, concolorous, glossy green. **Buds** in axils of leaves, in 7s to 13s, shortly stalked, broadly spindle-shaped, to 0.7 x 0.3 cm; operculum conical. **Flowers** creamy white, March to August. **Fruit** shortly stalked, cup-shaped to barrel-shaped, to 0.5 x 0.3 cm; rim sloping inwards, whitish.

DISTRIBUTION Widespread in mallee scrubs of southern Australia, north of the Nullarbor Plain, South Australia, Victoria and central New South Wales. Usually in sandy country.

NOTES A small mallee with fine glossy green leaves and dense clusters of buds and creamy white flowers.

Eucalyptus calycogona

SQUARE-FRUITED MALLEE

DESCRIPTION Mallee, often with steep branching habit. **Bark** smooth, grey to coppery, or rough at base. **Juvenile leaves** stalked, narrow-lanceolate to ovate, to 4 x 2.5 cm. **Adult leaves** stalked, narrow-lanceolate to lanceolate, to 10 x 1.5 cm, concolorous, glossy green. **Buds** in axils of leaves, in 7s, stalked, oblong, square in section, to 1.5 x 0.6 cm; operculum pyramidal. **Flowers** creamy white, July to November. **Fruit** stalked, oblong, square in section, to 1.5 x 0.8 cm.

DISTRIBUTION Widespread in the central and southern wheatbelts of Western Australia and the mallee scrubs of South Australia and Victoria, just extending into New South Wales.

NOTES Easily recognised in mallee scrub with its square buds and fruits, although it may grade into related species *E. gracilis* and *E. celastroides*.

Eucalyptus pimpiniana

PIMPIN MALLEE

DISTRIBUTION The Great Victoria Desert of South Australia with a small occurrence south of Lake Minigwal north-east of Kalgoorlie in Western Australia. Usually on sandy sites.

NOTES An attractive low shrubby mallee with greyish leaves, yellowish flowers and pendulous buds and fruits.

DESCRIPTION Low straggly, shrubby mallee. **Bark** smooth, white to pale coppery. **Juvenile leaves** stalked, ovate, to 11 x 7 cm. **Adult leaves** stalked, lanceolate to broad-lanceolate, to 11.5 x 3.3 cm, concolorous, dull, grey to blue-grey. **Buds** hanging down in axils of leaves, in 7s or more, stalked, cylindrical to elongated, to 2.8 x 1.1 cm; operculum beaked. **Flowers** yellowish. **Fruit** stalked, hanging down, cylindrical to barrel-shaped, to 2.2 x 1.3 cm.

Eucalyptus incrassata

RIDGE-FRUITED MALLEE

DESCRIPTION Mallee. **Bark** smooth, grey or grey-brown, or rough, loose and ribbony over most of stems. **Juvenile leaves** stalked, elliptical to ovate, to 10 x 5 cm. **Adult leaves** stalked, lanceolate to broad-lanceolate, to 11 x 3 cm, concolorous, glossy, green to olive green. **Buds** in axils of leaves, in 7s, stalked; base cup-shaped, operculum beaked, to 2.3 x 1 cm. **Flowers** creamy white, November to April. **Fruit** stalked, cup-shaped or cylindrical or slightly urn-shaped, smooth or ribbed, to 1.3 x 1.3 cm.

DISTRIBUTION Widespread in mallee scrubs of southern Australia, from north-east of Perth, east to south-western New South Wales, with a gap in the southern Nullarbor Plain. Usually confined to sandhills in South Australia, Victoria and New South Wales.

NOTES Apart from the red-fruited *E. forrestiana* group, it is one of the larger fruited mallees. It is closely related to *E. angulosa,* which is usually 3-flowered with larger buds and fruit and has a more coastal distribution.

Eucalyptus tetraptera

SQUARE-FRUITED MALLEE

DESCRIPTION Low straggly mallee. **Bark** smooth, grey or whitish grey. **Juvenile leaves** stalked, ovate to broadly elliptical, to 12 x 7 cm. **Adult leaves** stalked, elliptical to lanceolate, thick, to 25 x 7 cm, concolorous, glossy green. **Buds** rigidly downturned in axils of leaves, single, without a stalk, square in cross-section and almost winged along corners, bright red; base oblong, operculum pyramidal, to 5.5 x 3.4 cm. **Flowers** pink, August to December. **Fruit** without a stalk, oblong and winged along corners, to 5 x 4.2 cm.

DISTRIBUTION Restricted to coastal sandplains of Western Australia, from the Stirling Range National Park east towards Israelite Bay.

NOTES One of the most bizarre of all eucalypts with its spectacular, large, red buds and fruits. A popular, if straggly ornamental.

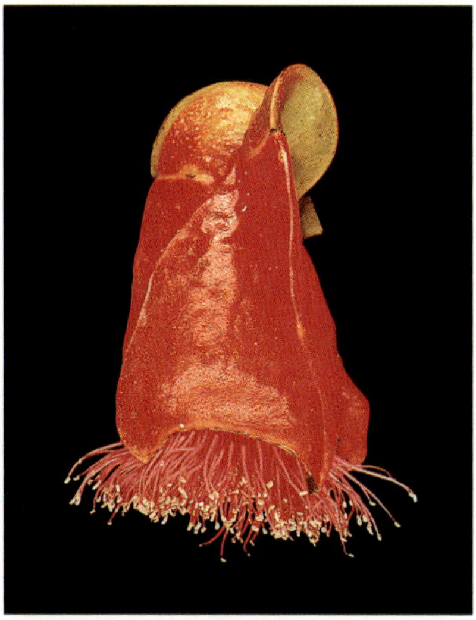

Eucalyptus stoatei

SCARLET PEAR GUM

DESCRIPTION Mallee or small tree. **Bark** smooth, grey, light grey or pale coppery. **Juvenile leaves** stalked, ovate, to 10 x 6 cm. **Adult leaves** stalked, oblong or elliptical to ovate, to 8 x 3 cm, concolorous, glossy green. **Buds** pendulous in axils of leaves, single, with stout stalk widening into coarsely ribbed base, to 3.5 x 1.9 cm; operculum hemispherical or shallowly conical. **Flowers** yellow, December to February. **Fruit** on stout stalks, barrel-shaped, ribbed, to 4 x 2.8 cm.

DISTRIBUTION Southern Western Australia, restricted to a small area east and north-east of Ravensthorpe to south of Pyramid Lake.

NOTES One of the most beautiful of the small trees or mallees with its spectacular large, red, ribbed, pendulous buds and fruits.

Eucalyptus forrestiana

FORRESTIANA MALLEE, FUCHSIA MALLEE

DESCRIPTION Mallee, mallet or small tree. **Bark** smooth, yellowish or greenish grey. **Juvenile leaves** stalked, ovate to lanceolate, to 10 x 6 cm. **Adult leaves** stalked, elliptical or oblong to lanceolate, to 10 x 2.3 cm, concolorous. **Buds** pendulous in axils of leaves, single, on tapering stalk, barrel-shaped in outline, square in cross-section and winged at the corners, to 6 x 2 cm (typical *forrestiana)* with pyramidal, flattened or hemispherical operculum; smaller in subspecies *dolichorrhyncha* and with a long beaked operculum. **Flowers** yellow, January to March. **Fruit** shortly, stoutly stalked, square in section and winged at corners, to 6 x 2.5 cm (*forrestiana*), to 4.5 x 2 cm (*dolichorrhyncha*).

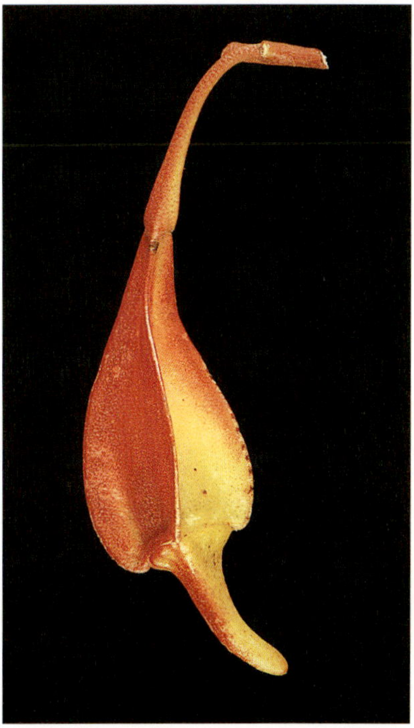

DISTRIBUTION South subcoastal Western Australia. Subspecies *forrestiana* occurs from north-east of Ravensthorpe east towards Mt Beaumont. The occurrence of subspecies *dolichorrhyncha* is much more restricted, north of Esperance towards Salmon Gums and north of the typical form.

NOTES The smaller and more delicate of the two subspecies (*dolichorrhyncha*) is one of the most popular ornamental eucalypts with its beautiful, large, pendulous, red buds and fruits. Its conspicuous presence and collections of seed along the main road north from Esperance have resulted in its early preference over subsp. *forrestiana* as an ornamental.

95

Eucalyptus concinna

VICTORIA DESERT MALLEE

DESCRIPTION Mallee. **Bark** rough on lower part of stems, flaky, grey-brown, or smooth, whitish grey to pale pink. **Juvenile leaves** stalked, lanceolate, to 8 x 3 cm. **Adult leaves** stalked, lanceolate to 10.5 x 1.7 cm, concolorous, very glossy green. **Buds** in axils of leaves, in 7s, stalked, cylindrical to ovoid with operculum wider than base, to 1 x 0.7 cm. **Flowers** white, November to March. **Fruit** stalked, cup-shaped to cylindrical, to 1 x 1 cm.

DISTRIBUTION Widespread from the goldfields of Western Australia eastwards through the Great Victoria Desert of South Australia.

NOTES An attractive mallee for the arid zones, with its extremely glossy leaves.

Eucalyptus torquata

CORAL GUM

DESCRIPTION Small tree or mallee. **Bark** rough on whole trunk or stems, hard, dark grey-brown to grey-black; branches smooth grey or coppery. **Juvenile leaves** stalked, elliptical to lanceolate, to 6 x 3 cm. **Adult leaves** stalked lanceolate, to 15 x 2.5 cm, concolorous, dull, grey-green. **Buds** pendulous in axils of leaves, in 7s, stalked; cylindrical but widened and ribbed at base, operculum strongly beaked, ribbed at base, to 3 x 1.1 cm. **Flowers** pink, August to November. **Fruit** stalked, cylindrical to urn-shaped, ribbed at base, to 1.7 x 1.2 cm.

DISTRIBUTION The goldfields of Western Australia from Coolgardie to south of Norseman and east of Kalgoorlie.

NOTES One of the most widely used ornamentals in southern Australia, easily recognised by its rough bark, grey-green leaves and pendulous buds and fruits with pink flowers.

Eucalyptus pterocarpa

RIB-FRUITED MALLEE

DESCRIPTION Small tree. **Bark** smooth, grey, whitish grey, salmon pink or coppery. **Juvenile leaves** stalked, ovate to triangular, to 14 x 8 cm. **Adult leaves** stalked, lanceolate, to 17 x 3 cm, concolorous, glossy green. **Buds** in axils of leaves, in 3s, stalked, diamond-shaped in outline, strongly ribbed, to 2.1 x 1.1 cm; operculum beaked. **Flowers** white, September to November. **Fruit** stalked, funnel-shaped, ribbed, to 1.8 x 1.7 cm.
DISTRIBUTION Western Australia, very restricted north-west of Norseman.
NOTES A handsome small tree, which has been widely cultivated since its relatively recent discovery in 1942. It was not formally described until 1988. It is popular for its large, beautifully sculptured buds and fruits.

Eucalyptus woodwardii

LEMON-FLOWERED MALLEE

DESCRIPTION Small tree or mallee. **Bark** smooth, white, pink, greenish or light coppery, shedding in ribbons. **Juvenile leaves** stalked, ovate to broad-lanceolate to elliptical, to 18 x 9 cm. **Adult leaves** stalked, broad-lanceolate, to 18 x 5 cm, concolorous, dull, grey-green to glaucous. **Buds** in axils of leaves, in 7s, stalked, base funnel-shaped to bell-shaped, operculum beaked, to 1.7 x 1.1 cm. **Flowers** lemon-yellow, August to November. **Fruit** stalked, bell-shaped, to 1.5 x 1.4 cm.

DISTRIBUTION Very restricted east of Kalgoorlie in Western Australia in the Karonie area.

NOTES A very popular ornamental in southern Australia because of its beautiful, large, lemon-yellow flowers.

Eucalyptus brevifolia

NORTHERN WHITE GUM

DESCRIPTION Small to medium-sized tree. **Bark** smooth, white, powdery. **Juvenile leaves** stalked, opposite, ovate, to 10 x 6 cm. **Adult leaves** stalked, lanceolate, to 10 x 2 cm, concolorous, dull, green to grey green. **Buds** in axils of leaves, in 7s, without stalks or shortly stalked, ovoid, to 0.8 x 0.5 cm; operculum obtusely conical. **Flowers** white, March to May. **Fruit** almost stalkless, cup-shaped, to 0.8 x 0.7 cm; rim thick.

DISTRIBUTION Widespread in the Kimberley of Western Australia, extending into the Northern Territory. Usually on low stony hills.

NOTES The common white-barked tree of stony hills occupying similar sites to *E. leucophloia,* which has thin-rimmed fruit and whose distribution is further to the south across Central Australia.

100

Eucalyptus leucophloia

WHITE GUM, MIGUM

DESCRIPTION Small tree. **Bark** smooth, white or pinkish, powdery. **Juvenile leaves** stalked, opposite, round to triangular, to 7 x 5 cm. **Adult leaves** stalked, lanceolate, to 10 x 2 cm, concolorous, dull, blue-grey to glaucous. **Buds** in axils of leaves, in 7s to 11s, shortly stalked, ovoid, to 0.9 x 0.5 cm; operculum hemispherical to conical. **Flowers** creamy white, April to June. **Fruit** shortly stalked, cupular to hemispherical, to 0.8 x 0.5 cm; valves below rim level (subsp. *leucophloia*) or strongly exserted (subsp. *euroa*).

DISTRIBUTION From the Pilbara in Western Australia (*leucophloia*) east across Central Australia to Cloncurry and Dajarra (*euroa*).

NOTES A beautiful white-barked tree of the arid zone, particularly stony hills. It has two forms distinguished as above by the fruiting valve characters.

Eucalyptus bigalerita

NORTHERN SALMON GUM

DESCRIPTION Small to medium-sized tree. **Bark** smooth, grey, pink, white or cream, becoming brilliant coppery. **Juvenile leaves** stalked, ovate to almost round, to 18 x 16 cm. **Adult leaves** stalked, ovate to triangular, to 15 x 9 cm, concolorous, glossy green. **Buds** in axils of leaves, in 7s, on short stout stalks, ovoid to almost spherical, to 1.2 x 1.1 cm; operculum hemispherical to very obtusely conical. **Flowers** white, June to September. **Fruit** crowded, without stalks or on short stalks, hemispherical to funnel-shaped, to 0.9 x 1.2 cm.

DISTRIBUTION North-western Australia, from the Kimberley north of the Drysdale River to the western part of the Top End of the Northern Territory north of Daly Waters. Usually on low seasonally wet flats.

NOTES One of the most beautiful trees of the north, with its spectacular coppery bark in season and the large glossy green leaves.

102

Eucalyptus apodophylla

WHITEBARK

DESCRIPTION Small to medium-sized tree. **Bark** smooth, pure white, or with yellow and pale orange patches. **Juvenile leaves** without stalks, opposite for many pairs, ovate, to 20 x 12 cm. **Leaves of mature crown** similar, ovate to oblong, to 15 x 7 cm, discolorous, dull, white waxy to bluish green or grey-green. **Buds** in axils of leaves, in 7s, rarely 3s, shortly stalked, ovoid or diamond-shaped, to 0.5 x 0.4 cm; operculum hemispherical or obtusely conical. **Flowers** white, July to October. **Fruit** shortly stalked, funnel-shaped, to 0.4 x 0.5 cm.

DISTRIBUTION North-western Australia, from the central and northern Kimberley east to the western part of the Top End of the Northern Territory. Usually growing in heavy soil in swamps and depressions. A rare form (subsp. *provecta*) with stalked leaves grows in the northern Kimberley.

NOTES A beautiful white-barked tree with greyish foliage.

Eucalyptus bancroftii

BANCROFT'S RED GUM, ORANGE GUM

DESCRIPTION Small to medium-sized tree. **Bark** smooth or granular, shedding over the whole trunk in large plates or flakes, newly exposed bark bright orange, weathering to grey and dark grey or grey-brown. **Juvenile leaves** stalked, ovate, to 12 x 7 cm. **Adult leaves** stalked, lanceolate to broad-lanceolate, to 20 x 4 cm, concolorous, green to grey-green. **Buds** in axils of leaves, in 7s to 11s, stalked, elongated, to 1.5 x 0.6 cm; operculum conical or horn-shaped, longer and narrower than base. **Flowers** white, November to January. **Fruit** stalked, hemispherical or funnel-shaped, to 0.9 x 0.9 cm, with a channel between rim and valves.

DISTRIBUTION Near-coastal distribution from north of Taree in New South Wales to near Maryborough in Queensland. Usually on infertile, wet sites.

NOTES The tableland form, *E. prava,* is usually glaucous.

Eucalyptus tereticornis

FOREST RED GUM, BLUE GUM, RED IRONGUM

DESCRIPTION Medium-sized to tall tree. **Bark** smooth or granular, shedding over whole trunk in large plates or flakes leaving patches of white, grey and bluish grey. **Juvenile leaves** stalked, ovate, to 21 x 10 cm. **Adult leaves** stalked, narrow-lanceolate to lanceolate, to 20 x 2.7 cm, concolorous, green, with numerous oil glands. **Buds** in axils of leaves, in 7s to 11s, stalked, elongated, to 2 x 0.5 cm; operculum horn-shaped, up to 7 times as long as the base. **Flowers** white, rarely pink, April to October. **Fruit** stalked, hemispherical (ovoid including the disc), to 0.6 x 0.8 cm; with rim and disc fused to base of exserted valves.

DISTRIBUTION Widespread along the coast and adjacent hills and plains of eastern Australia from south coastal New South Wales through Queensland, where it occurs inland as far as near Roma, Alpha, Charters Towers and Mt Surprise. It extends north to near Cooktown and also occurs in New Guinea.

NOTES Has the greatest latitudinal range of all eucalypts. Often with steeply ascending branches. Characterised by very long bud opercula and black, jagged seed.

Eucalyptus dealbata

TUMBLEDOWN RED GUM

DESCRIPTION Small to medium-sized tree. **Bark** smooth or granular, shedding in large plates or flakes exposing coppery, orange, white and grey colours. **Juvenile leaves** stalked, broad-ovate, to 14 x 8 cm, greyish green or glaucous. **Adult leaves** stalked, lanceolate, to 14 x 2.5 cm, concolorous, dull, grey-green, glaucous, with numerous oil glands. **Buds** in axils of leaves, in 7s or more, stalked, usually diamond-shaped, glaucous, to 1 x 0.5 cm; operculum conical or horn-shaped, longer than base. **Flowers** white, May to October. **Fruit** without stalks or shortly stalked, hemispherical, to 0.6 x 0.7 cm; valves exserted. Seed black, toothed.

DISTRIBUTION Occurs on the tablelands and western slopes of New South Wales from north of Canberra extending into south-eastern Queensland almost to Leyburn.

NOTES Generally of poor form. Glaucous grey-green leaves, twigs and buds are a feature.

106

Eucalyptus camaldulensis

RIVER RED GUM

DESCRIPTION Medium-sized to tall tree. **Bark** smooth, white, grey, yellow-green, grey-green or pinkish grey; sometimes with some rough, black bark accumulating at base. **Juvenile leaves** stalked, broad-lanceolate, to 26 x 8 cm, blue-green. **Adult leaves** stalked, lanceolate to narrow-lanceolate, to 18 x 2.2 cm, concolorous, dull or glossy, green or yellow-green to blue-green, with numerous oil glands. **Buds** in axils of leaves, in 7s to 11s or more, stalked, to 1 x 0.6 cm; base hemispherical, operculum beaked (var. *camaldulensis*) or obese (var. *obtusa*). **Flowers** white, July to February. **Fruit** stalked, ovoid (including the disc), to 0.7 x 1 cm; valves exserted. Seed yellow-brown, smooth.

DISTRIBUTION The var. *camaldulensis* is widespread along rivers, creeks and valleys of south-eastern Australia, particularly the Murray-Darling system, but including the coastal watersheds from lower Eyre Peninsula and Kangaroo Island to the Glenelg River. The var. *obtusa* is widespread along watercourses over a large part of mainland Australia, including the tropical north.

NOTES Probably the best known eucalypt, featuring in many paintings of rural areas. It is planted widely overseas for its timber and shade. It is a fast grower, and although naturally adapted to freshwater sites, it is also important for growing in saline situations.

107

Eucalyptus brassiana

CAPE YORK RED GUM, GUM-TOPPED PEPPERMINT, KARO (PAPUA NEW GUINEA)

DESCRIPTION Small to medium-sized tree or mallee. **Bark** rough, hard, dark grey to almost black for a few metres, smooth, whitish to yellowish above. **Juvenile leaves** stalked, broad-lanceolate to ovate, to 18 x 7 cm, green. **Adult leaves** stalked, lanceolate to narrow-lanceolate, to 18 x 2.5 cm, concolorous to slightly discolorous, slightly glossy, dark green. **Buds** in axils of leaves, in 3s or 7s, stalked, to 1.7 x 0.7 cm; base hemispherical, operculum long, conical. **Flowers** white, January to March. **Fruit** stalked, hemispherical (ovoid or rhomboidal including the disc), to 1.4 x 1.2 cm; valves exserted. **Seed** black, toothed.

DISTRIBUTION Of scattered occurrence on Cape York Peninsula north of a line between Cedar Bay and Aurukun, including Torres Strait Islands, with an isolated patch on 'Bullaringa' Station north-west of Mt Surprise. Also in New Guinea.

NOTES It has the largest and coarsest buds and fruit of the red gums, while the basal bark is like a shallow ironbark type.

Eucalyptus exserta

QUEENSLAND PEPPERMINT

DESCRIPTION Mallee or small to medium-sized tree. **Bark** rough and persistent on trunk and larger branches, moderately hard, with shallow longitudinal fissures, grey to dark grey; branches smooth, greyish. **Juvenile leaves** shortly stalked, linear, to 19 x 1.2 cm, slightly glossy, green or grey-green. **Adult leaves** stalked, lanceolate or narrow-lanceolate, to 18 x 1.7 cm, concolorous, slightly glossy, grey-green or dark green, with numerous oil glands. **Buds** in axils of leaves, in 7s, stalked, ovoid, to 1.3 x 0.5 cm; operculum conical, longer than base. **Flowers** white, November to February. **Fruit** stalkless or shortly stalked, hemispherical (ovoid including the disc), to 0.8 x 0.7 cm; valves exserted. **Seed** black, toothed.

DISTRIBUTION Central, eastern and south-eastern Queensland extending to west of Charleville and northwards to the Atherton Tableland; also north-east of Yetman in far northern New South Wales. Often on stony rises.

NOTES It belongs to a small group of red gums with some rough bark and very narrow juvenile leaves.

109

Eucalyptus ovata

SWAMP GUM

DESCRIPTION Small to medium-sized tree, rarely a mallee. **Bark** shedding over most of trunk to leave a smooth, grey, whitish or pinkish-grey surface; rough bark retained at base or for a few metres on larger trees. **Juvenile leaves** shortly stalked, ovate, to 19 x 8.5 cm, dull, green. **Adult leaves** stalked, broad-lanceolate, undulate, to 15 x 3 cm, concolorous, glossy, green. **Buds** in axils of leaves, in 7s, shortly stalked or occasionally without stalks, double conic, to 1.1 x 0.6 cm; operculum conical or slightly beaked. **Flowers** white, March to July. **Fruit** shortly stalked or occasionally without stalks, funnel-shaped, to 0.7 x 0.7 cm.

DISTRIBUTION The western end of Kangaroo Island, the southern Mount Lofty Range and the south-east of South Australia, Tasmania, the southern part of Victoria and south-eastern New South Wales. Prefers valleys and poorly drained flats.

NOTES Variable in habit from straggly saplings (in eastern Gippsland) to stout-boled trees. Leaves usually without visible oil glands, although these are present in the var. *grandiflora* of south-eastern South Australia and south-western Victoria.

Eucalyptus nicholii

NARROW-LEAVED BLACK PEPPERMINT,
WILLOW PEPPERMINT

DESCRIPTION Small to medium-sized tree. **Bark** rough and persistent to small branches, fibrous, coarsely fissured longitudinally, yellowish brown to grey-brown with red-brown underlayers. **Juvenile leaves** very shortly stalked, crowded, linear, to 5 x 0.5 cm, grey-green. **Adult leaves** stalked, narrow-lanceolate, to 13 x 1 cm, concolorous, dull, bluish green. **Buds** in axils of leaves, in 7s, stalked, spindle-shaped to ovoid, to 0.4 x 0.2 cm; operculum conical. **Flowers** white, March to April. **Fruit** stalked, funnel-shaped, hemispherical or somewhat bell-shaped, to 0.5 x 0.4 cm.

DISTRIBUTION Of limited occurrence on the northern tablelands of New South Wales, particularly in the area from Walcha to Tenterfield and to the east.

NOTES Widely cultivated as an ornamental in south-eastern Australia, the fine, dense, greyish foliage being particularly attractive.

Eucalyptus mannifera subsp. *mannifera*

BRITTLE GUM

DESCRIPTION Small to medium-sized tree. **Bark** shedding in irregular, greyish or pinkish plates or flakes to leave a smooth, white, powdery surface. Trees often with attractive pinkish red colouration of the older bark just before shedding. **Juvenile leaves** shortly stalked, linear to lanceolate or sickle-shaped, to 10 x 1.3 cm, blue-green to glaucous. **Adult leaves** stalked, lanceolate to narrow-lanceolate, to 15 x 1.5 cm, concolorous, dull, bluish green. **Buds** in axils of leaves, in 7s, shortly stalked, club-shaped or broadly spindle-shaped, to 0.5 x 0.3 cm; operculum conical or hemispherical. **Flowers** white, October to December. **Fruit** shortly stalked, hemispherical, to 0.5 x 0.4 cm.

DISTRIBUTION Occurs on the central and southern tablelands of New South Wales and adjacent parts of Victoria.

NOTES This species is favoured as an ornamental with its beautiful smooth white bark with seasonal red tints. There are several other subspecies: *E. mannifera* subsp. *praecox* (Capertee and Upper Cudgegong River districts of NSW, also north-west of Captains Flat); *E. mannifera* subsp. *elliptica* (western edges of the northern tablelands of NSW); *E. mannifera* subsp. *gullickii* (central and northern part of the NSW southern tablelands).

112

Eucalyptus scoparia

WALLANGARRA WHITE GUM

DESCRIPTION Small to medium-sized tree. **Bark** smooth throughout, white and light grey patches, with a powdery coating. **Juvenile leaves** stalkless or very shortly stalked, narrow-oblong, to 8 x 0.6 cm, glossy, green. **Adult leaves** stalked, narrow-lanceolate, to 15 x 1 cm, concolorous, glossy, green, with numerous oil glands. **Buds** in axils of leaves, in 7s, stalked, pear-shaped, to 0.5 x 0.3 cm; operculum conical or beaked, often slightly wider than base. **Flowers** white, November to December. **Fruit** stalked, hemispherical or cup-shaped (ovoid including the disc), to 0.5 x 0.5 cm.

DISTRIBUTION Of very restricted natural distribution in the Wallangarra region of southern Queensland, e.g. Christie Target, Castle Rock and Mt Norman. Also a very small patch on Mt Ferguson west of Amiens. Grows in clefts of massive granite outcrops on mountain tops.

NOTES Widely cultivated as an ornamental in south-eastern Australia for the attractive bark and fine, glossy green foliage, contrasting with that of *E. mannifera*.

Eucalyptus vernicosa

VARNISHED GUM

DESCRIPTION Dwarf, almost prostrate shrub, or small bushy tree. **Bark** smooth, greyish white, shedding in strips. **Juvenile leaves** sessile, opposite, elliptical or ovate, to 2.3 x 1.3 cm, glossy, green, with finely scalloped edges and distinctly pointed tips. **Stems** more or less square in cross-section. **Adult leaves** similar to the juveniles described, but shortly stalked and slightly larger, concolorous, glossy, green, thick, with conspicuous raised oil glands. **Buds** in axils of leaves, 3-flowered (or often apparently 1-flowered because of abortion of lateral buds), stalkless or with very short stalks, ovoid, to 0.8 x 0.5 cm; operculum conical. **Flowers** white, December to February. **Fruit** virtually without stalks, hemispherical to bell-shaped, to 0.8 x 0.5 cm.

DISTRIBUTION Endemic to Tasmania, occurring on high plateaus and mountain tops in the western part of the island.

NOTES Notable for its curious low shrubby habit and very small leaves.

Eucalyptus perriniana

SPINNING GUM

DESCRIPTION Shrub, mallee or small straggly tree. **Bark** mostly smooth, coppery, weathering to whitish green or light grey, or with a short basal stocking of thin, rough bark. **Juvenile leaves** connate (the bases of opposite leaves are joined around the stem), the pair together round or broadly elliptical, to 12 x 7 cm, grey-green, glaucous. **Adult leaves** stalked, lanceolate, to 12 x 2.5 cm, concolorous, grey-green or bluish green. **Buds** in axils of leaves, in 3s, stalkless or shortly stalked, ovoid, glaucous, to 0.7 x 0.4 cm; operculum hemispherical to shortly conical. **Flowers** white, January to March. **Fruit** sessile, hemispherical to cup-shaped, glaucous, to 0.7 x 0.7 cm.

DISTRIBUTION Of scattered distribution on high plateaus and mountains of south-eastern New South Wales, e.g. near Guthega and Kiandra and in the Tinderry Ranges; the Australian Capital Territory, e.g. Coronet Trig, Mt McKeahnie; eastern Victoria, e.g. the Dargo High Plains and Mt Donna Buang; and in Tasmania where it is restricted to a few localities at lower elevations, e.g. Hungry Flats, Tunnack and Strickland.

NOTES In the juvenile leaf phase the leaf tissue around the branchlets often dies, leaving the joined leaf pair free to spin in breezes.

Eucalyptus gunnii

CIDER GUM

DESCRIPTION Small to medium-sized tree. Older trees with a short, massive bole and large spreading branches. **Bark** often persistent for several metres as a thin, greyish stocking, or shedding all over to leave a smooth, yellowish, patchy surface, weathering to whitish, greenish and pinkish grey. **Juvenile leaves** stalkless, stem-clasping, opposite, ovate to round, with scalloped edges and notched ends, to 7 x 4 cm, grey-green, thick. **Adult leaves** stalked, elliptical to ovate, to 8 x 3 cm, concolorous, grey-green, thick. **Buds** in axils of leaves, in 3s, shortly stalked, usually glaucous, to 0.8 x 0.4 cm; base funnel-shaped, operculum flattish, conical or slightly beaked. **Flowers** white, January to February. **Fruit** stalkless, or very shortly stalked, often glaucous, cylindrical, sometimes with a slight constriction just below the rim, to 0.9 x 0.7 cm.

DISTRIBUTION Endemic to Tasmania, occurring on the plains and slopes of the central plateau to around 1100 m altitude, with an isolated occurrence south of Hobart.

NOTES This species is noted for exceptional cold tolerance and has been successfully cultivated in Britain and some parts of Europe. The related *E. archeri* is not glaucous (waxy).

Eucalyptus urnigera

URN GUM

DESCRIPTION Varying from a small to a tall tree. **Bark** smooth throughout except for some persistent, rough basal bark accumulating on larger trees, whitish, light grey, yellow-grey, yellow-green or reddish brown. **Juvenile leaves** stalkless, opposite for many pairs, stem-clasping, with scalloped edges and sometimes with notched ends, round, to 8 x 9 cm, glossy, green, or at higher altitudes somewhat glaucous. **Adult leaves** stalked, broad-lanceolate, to 9 x 2.5 cm, concolorous, glossy green or crown appearing slightly glaucous. **Buds** pendulous in axils of leaves, in 3s, on longish stalks, to 1.5 x 0.7 cm; base urn-shaped; operculum flattened and beaked. **Flowers** white, February to April. **Fruit** distinctly stalked, urn-shaped, often glaucous, to 1.6 x 1.1 cm.
DISTRIBUTION Endemic to Tasmania, occurring at high to moderately high altitudes, e.g. on Mount Wellington and in Mount Field National Park.
NOTES An attractive tree with its pendulous buds and fruits.

117

Eucalyptus cordata

HEART-LEAVED SILVER GUM

DESCRIPTION Shrub or small to medium-sized tree. **Bark** smooth throughout, white, green, purplish, grey or greenish yellow. **Juvenile leaves** stalkless, opposite, stem-clasping, with scalloped edges, round or heart-shaped, to 10 x 8 cm, greenish grey, glaucous; stems square in cross-section; shrubs and smaller trees mature and flower in the juvenile phase. **Adult leaves** (seen towards the top of larger trees) stalked, lanceolate to broad-lanceolate, to 13 x 3.5 cm, concolorous, grey-green to glaucous. **Buds** in axils of leaves, in 3s, stalkless or very shortly stalked, glaucous, to 1 x 0.6 cm; base cylindrical; operculum flattened and beaked. **Flowers** white, August to September. **Fruit** stalkless, cup-shaped to cylindrical, glaucous, to 1.3 x 1.3 cm.

DISTRIBUTION Endemic to Tasmania, of restricted distribution in the south-east at intermediate altitudes, e.g. the foothills of Mount Wellington, Snug Plains, Port Arthur and Moogara.

NOTES An attractive ornamental with its large, roundish, glaucous juvenile leaves, which often persist in the crown.

Eucalyptus pulverulenta

SILVER-LEAVED MOUNTAIN GUM

DESCRIPTION Straggly shrub, mallee or small tree. **Bark** smooth, shedding to ground level but often with ribbons of older bark hanging from the stems. Young stems intensely glaucous with a white wax. **Juvenile leaves** stalkless, opposite, stem-clasping, round to ovate, to 5 x 5 cm, bluish grey, glaucous. **Plants** are mature in the juvenile phase with adult leaves very rarely formed; shortly stalked, oblong to lanceolate, to 10 x 2 cm, concolorous, bluish grey, glaucous. **Buds** in axils of leaves, in 3s, stalkless, ovoid or diamond-shaped, intensely glaucous, to 0.8 x 0.4 cm; operculum conical. **Flowers** white, September to November. **Fruit** stalkless, hemispherical to cup-shaped, glaucous, to 0.8 x 0.8 cm.

DISTRIBUTION Endemic to south-east New South Wales, occurring in disjunct populations: the upper Cox's River area south-west of the Blue Mountains, the southern highlands east of Bredbo and west of Bombala.

NOTES While very restricted in natural populations this species has been widely cultivated as an ornamental for its unusual and attractive foliage. Differs from the similar *E. cinerea* by the smooth bark and largely mallee habit.

119

Eucalyptus cinerea

ARGYLE APPLE, MEALY STRINGYBARK

DESCRIPTION Small to medium-sized tree. **Bark** rough and persistent on the trunk and larger branches, thick, fibrous, longitudinally furrowed, reddish brown to grey-brown. **Juvenile leaves** stalkless, stem-clasping, opposite, round to heart-shaped, to 8 x 5.5 cm, greyish blue, glaucous. **Trees** are usually mature in the juvenile phase but often produce intermediate and adult leaves which are stalked, broad-lanceolate, to 11 x 2 cm, concolorous, greyish blue, glaucous. **Buds** in axils of leaves, in 3s, stalkless or shortly stalked, glaucous, diamond-shaped, to 0.7 x 0.4 cm; operculum conical. **Flowers** white, October to December. **Fruit** stalkless or shortly stalked, funnel-shaped, to 0.8 x 0.9 cm.

DISTRIBUTION Typically from north of Bathurst, in central west New South Wales, to the Beechworth area of northern Victoria.

NOTES A related form with a greater proportion of the crown with adult leaves, subsp. *triplex*, occurs on rocky mountain sites south of Canberra and in low hills north-west of Captains Flat.

Eucalyptus angophoroides

APPLE-TOPPED BOX

DESCRIPTION Medium-sized to tall tree. **Bark** rough and persistent over trunk and larger branches, fibrous, flaky, with shallow fissures, grey and often with whitish bleached patches. **Juvenile leaves** shortly stalked, stem-clasping, with scalloped edges, ovate to heart-shaped, to 9 x 7 cm, strongly discolorous, dark green above, paler below. **Adult leaves** stalked, lanceolate to narrow-lanceolate, to 20 x 2.5 cm, distinctly or slightly discolorous, dark green above. **Buds** in axils of leaves, in 7s, stalked, ovoid, to 0.7 x 0.5 cm; operculum conical or beaked. **Flowers** white, October to December. **Fruit** stalked, funnel-shaped or hemispherical, to 0.5 x 0.7 cm.

DISTRIBUTION Occurs on the coastal plains and adjoining escarpments of the south coast of New South Wales from Batemans Bay south and south-west to the Strzelecki Range in the Gippsland region of Victoria.

NOTES It differs from the related *E. bridgesiana* by the green juvenile leaves, slightly discolorous adult leaves and fruit with 4 valves (mostly 3 in *E. bridgesiana*).

Eucalyptus globulus

TASMANIAN BLUE GUM, SOUTHERN BLUE GUM

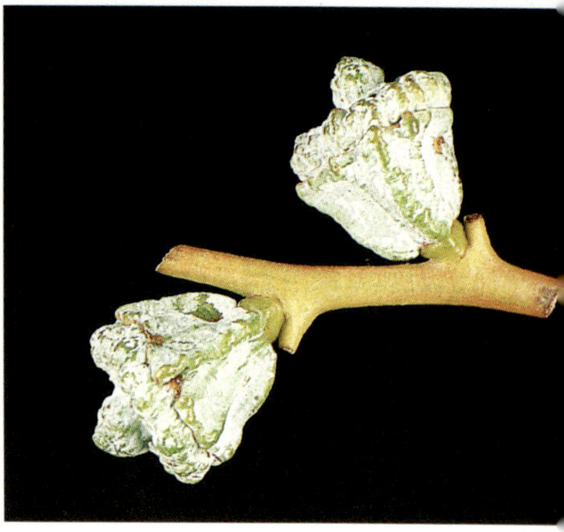

DESCRIPTION Medium-sized to very tall tree. **Bark** smooth throughout, shedding in strips and patches leaving creamy white, light and dark grey and yellowish colours. **Juvenile leaves** stalkless, opposite, stem-clasping, ovate, to 15 x 10.5 cm, bluish green, glaucous; stems square and flanged in cross-section, very glaucous. **Adult leaves** stalked, usually sickle-shaped, also lanceolate to narrow-lanceolate, to 30 x 3 cm, glossy, green. **Buds** in axils of leaves, mostly 1-flowered (but occasionally in 3s), usually stalkless, occasionally very shortly stalked, very glaucous, to 2.3 x 1.8 cm, more or less top-shaped with 4 (or occasionally more) ribs on base; operculum flattened with a distinct central knob, very warty. **Flowers** white, September to December. **Fruit** stalkless, subglobular to more or less hemispherical, to 2.1 x 2.5 cm; base glaucous with 4 or more distinct ribs; disc broad.

DISTRIBUTION Occurs extensively in eastern and south-eastern Tasmania, with isolated patches elsewhere, including the west coast; on King and Flinders Islands in Bass Strait; and from the Otway Ranges area to the south-west of Melbourne, Wilsons Promontory and areas immediately north.

NOTES Part of the southern blue gum group of closely related species—*E. maidenii*, *E. pseudoglobulus* and *E. bicostata*, occurring in Victoria and New South Wales. Easily recognised by the single flowers in the axils (rarely in 3s). The four forms are usually referred to as subspecies.

122

Eucalyptus nitens

SHINING GUM, SILVERTOP

DESCRIPTION Tall to very tall tree. **Bark** smooth or rough and persistent at the base, thin, flaky, dark grey to black, shedding above in large strips to leave a smooth yellowish, grey or white surface. **Juvenile leaves** stalkless, opposite, stem-clasping, broad-lanceolate to ovate, to 17 x 8 cm, discolorous, greyish blue; stems squarish in cross-section, flanged and glaucous. **Adult leaves** stalked, lanceolate, to 25 x 2.5 cm, concolorous, glossy, green. **Buds** in axils of leaves, in 7s, stalkless, ovoid, to 0.7 x 0.3 cm; operculum conical. **Flowers** white, January to March. **Fruit** stalkless, cup-shaped or barrel-shaped, to 0.7 x 0.6 cm, with a glossy surface as though varnished.

DISTRIBUTION Occurs on high tablelands and mountains of coastal ranges of southern New South Wales, e.g. Tallaganda State Forest and Brown Mountain, with two small, disjunct, very high altitude populations at Barrington Tops east of Scone and near Ebor. Also mountains of eastern Victoria.

NOTES Populations in Bondi State Forest and the Errinundra Plateau in the New South Wales–Victorian border region once known as *E. nitens* have been recently described as *E. denticulata*. The adult leaves have very distinctive toothed margins.

Eucalyptus cypellocarpa

MOUNTAIN GREY GUM

DESCRIPTION Medium-sized to very tall tree. **Bark** shedding to ground level, smooth, yellow-white or greyish. **Juvenile leaves** stalkless, opposite, broad-lanceolate to elliptical or ovate, to 17 x 7 cm, glossy, green. **Adult leaves** stalked, lanceolate or sickle-shaped, to 20 x 2.6 cm, concolorous, green. **Buds** in axils of leaves, in 7s, stoutly stalked, ovoid, to 1.2 x 0.5 cm, often ribbed; operculum beaked or conical, much shorter than base. **Flowers** white, December to February. **Fruit** stalked, cup-shaped, cylindrical or barrel-shaped, to 1 x 0.9 cm.

DISTRIBUTION Abundant in the southern tablelands and coastal foothills of New South Wales and central and eastern Victoria; also in the Otway Ranges and the Grampians of Victoria.

NOTES One of a variable group of species, two of which, *E. volcanica* and *E. retinens*, are largely rough-barked and occur in northern New South Wales. The Grampians form includes the slightly different *E. alaticaulis*.

Eucalyptus goniocalyx

LONG-LEAVED BOX

DESCRIPTION Small to medium-sized tree. **Bark** rough and persistent to the small branches, fibrous, greyish, becoming deeply fissured, thick and shaggy in larger trees. **Juvenile leaves** stalkless, opposite, round, to 11 x 10 cm, light green to glaucous. **Adult leaves** stalked, lanceolate, to 20 x 3 cm, concolorous, green. **Buds** in axils of leaves, in 7s, stalkless, ovoid, to 1.3 x 0.6 cm, base usually ribbed; operculum conical. **Flowers** white, March to July. **Fruit** stalkless, cup-shaped or cylindrical, sometimes ribbed, to 1 x 1 cm.

DISTRIBUTION Occurs widely on the tablelands of New South Wales and in Victoria west almost to the South Australian border. Also in South Australia from the Mount Lofty Ranges to the Flinders Ranges.

NOTES It is related to the dull-leaved, poorly formed *E. nortonii* of southern New South Wales and central and eastern Victoria. A little known mallee form with smooth or some rough bark occurs in the higher parts of the Northern Flinders Range, e.g. St Marys Peak.

Eucalyptus viminalis

MANNA GUM, RIBBON GUM, WHITE GUM

DESCRIPTION Medium-sized to very tall tree. **Bark** mostly shedding to ground level, smooth, grey or whitish; some trees retain a basal stocking of dark grey-brown to black rough bark. **Long ribbons** of partly shed bark can usually be seen in tree crowns. **Juvenile leaves** stalkless, opposite, stem-clasping, lanceolate, to 15 x 3 cm, green. **Adult leaves** stalked, lanceolate to narrow-lanceolate, often undulate, to 20 x 2 cm, concolorous, green. **Buds** in axils of leaves, in 3s, stalkless or shortly stalked, ovoid to broadly spindle-shaped, to 1 x 0.5 cm; operculum pointed-hemispherical or conical. **Flowers** white, January to May. **Fruit** stalkless to shortly stalked, cup-shaped or hemispherical, to 0.8 x 0.9 cm.

DISTRIBUTION Widespread on tablelands from the southern part of the northern tablelands of New South Wales, through most of southern Victoria, eastern Tasmania and Mt Lofty Range and south-east of South Australia. Prefers valley situations but also coastal in Tasmania and south of Eden in New South Wales.

NOTES *E. viminalis* subsp. *cygnetensis,* occurs in parts of South Australia—Kangaroo Island, south of Port Lincoln, Mount Lofty Ranges, southern Murray Mallee and the south-east corner—extending into adjacent parts of Victoria. *E. nobilis* (formerly regarded as a 7-flowered form of *E. viminalis*) occurs in the north-eastern parts of the northern tablelands of New South Wales, extending into southern Queensland to around Cunninghams Gap.

126

Eucalyptus crenulata

SILVER GUM, BUXTON GUM

DESCRIPTION Small tree. **Bark** rough on older trees over half or more of the trunk, moderately thin, compact, hard, light grey; shedding above to leave a smooth, yellowish, grey or grey-brown surface. **Juvenile leaves** stalkless, stem-clasping, ovate and heart-shaped, with distinctly scalloped edges, to 4 x 2 cm, discolorous, blue-green, glaucous. **Branchlets** extremely glaucous with a coating of whitish wax. **Adult leaves** similar to the juveniles described, some pairs becoming sub-opposite and shortly stalked; older leaves inside crown glossy green. **Buds** in axils of leaves, in 7s to 15s, stalked, ovoid or club-shaped, very glaucous, to 0.5 x 0.3 cm; operculum beaked. **Flowers** white, September to October. **Fruit** in globular clusters, shortly stalked, cup-shaped, very glaucous, to 0.4 x 0.4 cm.

DISTRIBUTION Very restricted natural distribution north-east of Melbourne in the valley of the Acheron River, near Buxton and Yering; on swampy sites.

NOTES Has been widely and successfully cultivated in southern Australia for the attractive foliage and relatively small stature.

Eucalyptus neglecta

OMEO GUM

DESCRIPTION Small tree. **Bark** rough on the trunk, greyish, fibrous, held in elongated strips formed by shallow longitudinal furrows, smooth, greenish or greenish grey above. **Juvenile leaves** stalkless, opposite, broadly elliptical to round, to 11 x 7 cm, grey-green; glaucous on the mid-ribs and with stems square in cross-section. **Adult leaves** stalked, lanceolate to broad-lanceolate, to 15 x 3.5 cm, concolorous, green. **Buds** in axils of leaves, in 7s to 15s, stalkless, ovoid, to 0.4 x 0.3 cm; outer operculum shedding imperfectly, reddish brown and contrasting with the glaucous base; inner operculum hemispherical or shallowly conical. **Flowers** white, December to January. **Fruit** in more or less globular clusters, stalkless, hemispherical or slightly funnel-shaped, to 0.6 x 0.5 cm.

DISTRIBUTION Of scattered and restricted distribution in mountains of eastern Victoria, e.g. Livingstone and Spring Creeks near Omeo, the upper Buckland Valley, the Dargo High Plains and Frenchs Creek east of Marysville; usually an understorey species near perennial streams.

NOTES Not widely known but is an attractive ornamental with its small form and large leaves.

Eucalyptus kitsoniana

GIPPSLAND MALLEE

DESCRIPTION Mallee or small to medium-sized tree. **Bark** smooth, yellow at first, weathering to white or ashy grey; older trees may have several metres of thin, brownish rough bark. **Juvenile leaves** stalkless, opposite, ovate to round, to 10 x 8 cm, light green. **Adult leaves** stalked, lanceolate to broad-lanceolate, to 18 x 3 cm, concolorous, green, relatively thick. **Buds** in axils of leaves, in 7s, stalkless, broadly ovoid, to 0.9 x 0.4 cm, with conspicuous bracts around whole bud cluster when immature; operculum broadly conical. **Flowers** white, January to February. **Fruit** stalkless, cup-shaped to hemispherical, to 0.8 x 0.8 cm.

DISTRIBUTION Of restricted and strictly coastal occurrence in Victoria, east and west of Melbourne, e.g. South Gippsland, Cape Nelson, lower Glenelg River, Cape Otway.

NOTES A useful ornamental especially for exposed coastal areas.

Eucalyptus sturgissiana

ETTREMA MALLEE

DESCRIPTION Slender-stemmed mallee. **Bark** smooth, with glaucous first- and second-year branchlets, becoming grey, light grey-brown, pinkish or coppery. **Juvenile leaves** stalkless, opposite, at first elliptical then round and finally ovate and connate (the pair joined and encircling the stem), to 7 x 6 cm, light green, with glaucous midribs and stems. **Adult leaves** with short, flat stalks, lanceolate to narrow-lanceolate, often with hooked tips, to 12 x 1.5 cm, concolorous, light green. **Buds** in axils of leaves, in 7s, stalkless or very shortly stalked, usually glaucous, broadly spindle-shaped, to 0.7 x 0.5 cm; operculum conical or hemispherical with a small point. **Flowers** white, August to November. **Fruit** stalkless or very shortly stalked, hemispherical or cup-shaped, to 0.6 x 0.7 cm.

DISTRIBUTION Of very restricted distribution in New South Wales on the sandstone plateaus south-west of Nowra, with a few small occurrences nearby on the coastal plain.

NOTES A relatively recently discovered species, which has possibilities as an attractive ornamental with its straggly low form and curious, connate juvenile leaves.

Eucalyptus leptophleba

MOLLOY RED BOX

DESCRIPTION Small to medium-sized tree. **Bark** rough to small branches, with irregular tessellations, greyish with lighter patches. **Juvenile leaves** stalked, ovate, to 15 x 9 cm, greyish green. **Adult leaves** stalked, lanceolate, to 24 x 3 cm, concolorous, dull, bluish green or light green. **Bud clusters** in axils of leaves and at ends of branchlets, in 7s to 11s. **Buds** stalked, club-shaped to ovoid, to 1.2 x 0.6 cm; operculum hemispherical to conical. **Flowers** white, March to August. **Fruit** stalked, cup-shaped to truncate-ovoid, to 1.1 x 1 cm.

DISTRIBUTION Widespread in far north Queensland from north of Hughenden and Cairns, including the Atherton Tableland and most of Cape York Peninsula except the south-west, and an isolated occurrence south of Cardwell. Also on Thursday Island and in New Guinea.

NOTES One of the many northern boxes, it is the only species of the group to attain considerable tree size in far north Queensland.

Eucalyptus patellaris

WEEPING BOX

DESCRIPTION Small to medium-sized tree. **Bark** rough to small branches, box-type, grey over dark brown. **Juvenile leaves** stalked, elliptical or ovate or more or less triangular, to 15 x 4 cm. **Adult leaves** stalked, narrow-lanceolate often curved, to 23 x 2 cm, concolorous, glossy, green. **Bud clusters** in axils of leaves and at ends of branchlets, in 3s to 11s. **Buds** stalked, pear-shaped to diamond-shaped, to 1.5 x 0.8 cm; operculum conical to cap-shaped (wider than base at join). **Flowers** white, November to February. **Fruit** stalked, cup-shaped to bell-shaped, to 1.1 x 1 cm, with prominently flared rim; valves exserted.

DISTRIBUTION Scattered distribution in the Top End of the Northern Territory, usually on alluvial flats and in the vicinity of watercourses.

NOTES A northern box easily recognised by the habitat, glossy green leaves and large, flared fruit.

132

Eucalyptus chlorophylla

GLOSSY-LEAVED BOX

DESCRIPTION Small to medium-sized tree or mallee. **Bark** rough to the small branches, box-type, brown, grey or whitish grey. **Juvenile leaves** stalked, lanceolate to broad-lanceolate, to 9 x 2.3 cm, glossy, green. **Adult leaves** stalked, narrow-lanceolate or curved, to 18 x 2 cm, concolorous, glossy, green to dark green. **Bud clusters** in axils of leaves and at ends of branchlets, in 7s. **Buds** stalked, broadly spindle-shaped to club-shaped, to 0.7 x 0.3 cm; operculum conical. **Flowers** white, November to January. **Fruit** stalked, funnel-shaped or bell-shaped, to 0.8 x 0.5 cm.

DISTRIBUTION Scattered but geographically widespread distribution from east of Kununurra in Western Australia across the Northern Territory north of Tennant Creek, to east of Normanton in north-west Queensland and on Cape York Peninsula.

NOTES A northern box easily recognised by the small form and bright glossy green crown.

Eucalyptus pruinosa

SILVER BOX

DESCRIPTION Small tree or mallee. **Bark** rough to small branches, box-type, often in longitudinal strips, grey over dark brown. **Juvenile leaves** shortly stalked or stalkless, opposite, stem-clasping, elliptical to ovate, to 10 x 2.5 cm, dull, bluish, glaucous. **Adult leaves** stalkless, opposite, often stem-clasping, broadly elliptical to ovate, to 14 x 9 cm, concolorous, dull, bluish, glaucous; or older leaves inside crown can be glossy, green. **Bud clusters** in axils of leaves and at ends of branchlets, in 3s and 7s. **Buds** stalked, broadly spindle-shaped, usually glaucous, to 1 x 0.5 cm; operculum conical to beaked. **Flowers** white, November to January. **Fruit** stalked, funnel-shaped, usually glaucous, to 1 x 0.7 cm.

DISTRIBUTION Widespread and very common in northern Australia from the central Kimberley region of Western Australia, across the central Northern Territory extending into north-western Queensland to Mt Isa and Croydon.

NOTES The most conspicuous of the northern boxes with large glaucous foliage. Occurs in very large stands over vast areas.

Eucalyptus microtheca

COOLIBAH

DESCRIPTION Small to medium-sized tree or mallee. **Bark** rough to the small limbs, subfibrous, light or dark grey. **Juvenile leaves** stalked, lanceolate to narrow-lanceolate, to 15 x 2.5 cm, blue-green to greyish. **Adult leaves** stalked, lanceolate, to 15 x 2.5 cm, concolorous, dull, blue-green to grey or glaucous. **Bud clusters** in axils of leaves and at ends of branchlets, in 7s. **Buds** stalked, ovoid or pear-shaped, often glaucous, to 0.3 x 0.3 cm; operculum conical to slightly beaked. **Flowers** white, November to December. **Fruit** stalked, funnel-shaped or hemispherical, to 0.4 x 0.4 cm.

DISTRIBUTION Widespread in northern Australia from the Ord River flats of the Kimberley region of Western Australia, to the Top End of the Northern Territory from Daly Waters northwards, extending into north-western Queensland, east to the Gilbert River and south-western Cape York Peninsula. On heavy soil associated with rivers, creeks and flood plains.

NOTES The other well-known coolibah (*E. coolabah)* is widespread across Australia from the Darling River and Lake Eyre drainage systems extending through much of Queensland and the Northern Territory and across to the Fitzroy River in the southern Kimberley. This species is very similar to *E. microtheca* in most respects but has conspicuous, smooth, whitish branches and more exserted valves of the fruit.

135

Eucalyptus populnea

POPLAR BOX, BIMBLE BOX

DESCRIPTION Small to medium-sized tree. **Bark** rough and persistent on trunk and larger branches, box-type, light grey to grey-brown, smaller branches often smooth, whitish. **Juvenile leaves** stalked, ovate to round, often broader than long, to 11.5 x 9.5 cm, glossy, green, thick. **Adult leaves** stalked, broad-lanceolate to ovate or round, to 11 x 3 cm, concolorous, glossy, green, with numerous oil glands. **Bud clusters** in axils of leaves and at ends of branchlets, in 7s to 15s or more. **Buds** shortly stalked, club-shaped or ovoid to sub-globular, to 0.5 x 0.3 cm; operculum conical to hemispherical. **Flowers** white, February to March. **Fruit** shortly stalked, funnel-shaped to hemispherical, to 0.4 x 0.5 cm.

DISTRIBUTION Widespread and very abundant in Queensland from west of Mackay to Barcaldine, south to Hungerford and east to Warwick; also on the western plains of New South Wales and west of the Darling River.

NOTES Easily recognised by the large round glossy green leaves. In the northern region of its distribution *E. populnea* grades into the closely related, narrow-leaved *E. brownii,* which is common between Townsville and Hughenden.

Eucalyptus behriana

BULL MALLEE, BROAD-LEAVED BOX

DESCRIPTION Small to tall mallee. **Bark** rough on part or whole of trunk, box-type, dark brown to brown-black; smooth, grey or pinkish grey over cream or greenish above. **Juvenile leaves** stalked, ovate or +/– triangular, to 12 x 7 cm, light green and slightly glossy, or blue-green to glaucous. **Adult leaves** stalked, broad-lanceolate to ovate, to 10 x 3 cm, concolorous, glossy, green. **Bud clusters** at ends of branchlets, in 3s and 7s. **Buds** stalkless or very shortly stalked, club-shaped or ovoid, to 0.7 x 0.3 cm; operculum conical to hemispherical. **Flowers** white, November to January. **Fruit** usually stalkless, cup-shaped to barrel-shaped, to 0.6 x 0.5 cm.

DISTRIBUTION Of scattered and disjunct distribution on the plains of the mid-north of South Australia, e.g. Riverton and Nuriootpa, the northern part of the Mount Lofty Ranges, and on lower Eyre Peninsula, e.g. south of Cummins; also around Wyalong in central New South Wales; and near Bacchus Marsh and in north-western Victoria extending to the Bordertown district in South Australia.

NOTES Curiously scattered but remarkably uniform apart from the juvenile leaf colour. The young bud clusters are conspicuous with their completely leafless stems and oblong, stubby buds.

Eucalyptus lansdowneana

RED-FLOWERED MALLEE BOX

DESCRIPTION Slender-stemmed, straggly mallee. **Bark** smooth, grey over creamy white. **Juvenile leaves** stalked, ovate to broad-lanceolate, to 10 x 6 cm, glossy, green. **Adult leaves** stalked, lanceolate to broad-lanceolate, to 15 x 3 cm, glossy, green to yellow-green. **Bud clusters** in axils of leaves and at ends of branchlets, in 7s. **Buds** without stalks or very shortly stalked, ovoid to club-shaped, to 1.1 x 0.6 cm; operculum conical. **Flowers** red and pinkish red, August to October. **Fruit** stalkless, cup-shaped to funnel-shaped, to 1.2 x 1.1 cm.

DISTRIBUTION Of very restricted natural occurrence, known only from a few rocky hills in the Gawler Range, upper Eyre Peninsula, South Australia.

NOTES The name *lansdowneana* was incorrectly applied to another species of lower Eyre Peninsula and Kangaroo Island; this is the unrelated box species, *E. lansdowneana* subsp. *albopurpurea,* with white, pink or mauve flowers and which is grown widely as an ornamental, particularly in Perth. True *E. lansdowneana* is a beautiful, slender mallee with large, glossy leaves and coloured flowers.

Eucalyptus intertexta

GUM-BARKED COOLIBAH, COOLIBAH, GUM COOLIBAH, SMOOTH-BARKED COOLIBAH

DESCRIPTION Small to medium-sized tree, rarely a mallee. **Bark** variable, rough and persistent on lower part or whole of trunk, thin or moderately thick and shaggy, grey or red-brown; smooth, whitish or greyish white above. **Juvenile leaves** stalked, elliptical to broad-lanceolate then ovate, to 12 x 5 cm, glaucous or bluish green to greyish green. **Adult leaves** stalked, narrow to broad-lanceolate, to 14 x 2.5 cm, concolorous, dull, grey-green to blue-grey. **Bud clusters** mostly at ends of branchlets with some branched and occasionally unbranched groups in axils of leaves, in 7s. **Buds** stalked, club-shaped, to 1 x 0.4 cm; operculum conical. **Flowers** white, March to September. **Fruit** stalked, cup-shaped to barrel-shaped, to 0.8 x 0.7 cm.

DISTRIBUTION Geographically widespread but of very scattered and disjunct occurrence; in Queensland from north-west of Dirranbandi, to west of Charleville and north of Augathella, with a smaller occurrence north of Talwood; also arid parts of South Australia, north of the River Murray, the Northern Flinders Range, northern Eyre Peninsula and further inland; in the eastern part of the Great Victoria Desert in Western Australia (south-west of Warburton) and extending into the southern part of Northern Territory; also on the western plains of New South Wales.

NOTES One of the most widespread species of the arid zones, it occurs on plains and hills, and in some situations is of mallee form, e.g. in the Gibson Desert and the Everard Range. The latter is probably a distinctive subspecies.

139

Eucalyptus cambageana

DAWSON GUM, DAWSON RIVER BLACKBUTT

DESCRIPTION Small to medium-sized tree. **Bark** rough, thick on lower trunk, tessellated, dark brown to black; smooth, pinkish grey to white and yellowish above. **Juvenile leaves** stalked, ovate, to 14.5 x 8.5 cm, bluish. **Adult leaves** stalked, lanceolate to narrow-lanceolate, to 17 x 2 cm, concolorous, slightly glossy, green. **Bud clusters** mostly at ends of branchlets, in 7s. **Buds** stalked, club-shaped to broadly spindle-shaped, to 0.7 x 0.4 cm; operculum hemispherical to conical. **Flowers** white, December to January. **Fruit** stalked, funnel-shaped to cup-shaped, to 0.7 x 0.5 cm.

DISTRIBUTION Mainly in central-eastern Queensland from around Charters Towers south almost to Tambo and east to Biloela; also north and west of Charleville, in the Taroom district and south of Rockhampton.

NOTES Easily recognised by the black butt, glossy green leaves, and fruit with a whitish, inward-sloping disc.

Eucalyptus largiflorens

BLACK BOX

DESCRIPTION Small to medium-sized tree. **Bark** rough and persistent to the smallest branches, dark grey, becoming thick and shaggy in old trees. **Juvenile leaves** shortly stalked, linear, to 15 x 1 cm, bluish. **Adult leaves** stalked, narrow-lanceolate to lanceolate, to 18 x 1.8 cm, concolorous, dull, grey-green. **Bud clusters** mostly at ends of branchlets, in 7s to 11s. **Buds** shortly stalked, ovoid, to 0.5 x 0.5 cm; operculum conical to hemispherical. **Flowers** white (rarely pink), August to January. **Fruit** stalked, hemispherical to cup-shaped, to 0.6 x 0.5 cm.

DISTRIBUTION Widespread in western New South Wales, extending into southern Queensland north of Goondiwindi and along the Moonie River; in north-western Victoria and into South Australia along the River Murray as far as Mannum (rarely further south), with some populations west of the river and the northern Mount Lofty Ranges towards Gulf St Vincent. This species prefers seasonally inundated, heavy-soiled river flats and lake edges.

NOTES A southern box, usually recognised by the habitat, rough bark, dull, narrow leaves, and particularly by the long, narrow juvenile leaves of coppice and seedlings.

Eucalyptus albens

WHITE BOX

DESCRIPTION Small to medium-sized tree. **Bark** rough and persistent over the whole trunk and larger branches, greyish; branches above this smooth, greyish or white. **Juvenile leaves** stalked, ovate to round, to 15 x 11.5 cm, bluish grey, glaucous. **Adult leaves** stalked, broad-lanceolate, to 15 x 3 cm, concolorous, pale slate grey to bluish grey. **Bud clusters** in axils of leaves and at ends of branchlets, in 7s. **Buds** stalkless or stalked, broadly spindle-shaped, usually angular, often somewhat curved, usually glaucous, to 1.8 x 0.6 cm; operculum conical to beaked. **Flowers** white, March to June. **Fruit** varies from stalkless to distinctly stalked, barrel-shaped to slightly urn-shaped, often with angles or ribs, usually glaucous, to 1.3 x 1 cm.

DISTRIBUTION Extends from the south-east corner of Queensland, from north-north-east of Dalby, throughout the western slopes and tablelands of New South Wales, to the upper Snowy River region in eastern Victoria, with minor occurrences further west in that State, and an isolated, disjunct occurrence around Mt Remarkable in the South Flinders Range in South Australia.

NOTES A southern box recognised by the smooth branches and rather broad, greyish leaves. Buds and fruit are large for a box species.

Eucalyptus moluccana

GREY BOX, GUM-TOPPED BOX

DESCRIPTION Medium-sized to occasionally tall tree. **Bark** rough and persistent on lower part or most of trunk, shedding above to leave a smooth, whitish or light grey, sometimes shiny surface. **Juvenile leaves** stalked, ovate to round, to 17 x 10 cm, glossy, green. **Adult leaves** stalked, broad-lanceolate to lanceolate, to 14 x 3.5 cm, concolorous, glossy, green, with numerous oil glands. **Bud clusters** in axils of leaves and at ends of branchlets, in 7s. **Buds** stalked, broadly spindle-shaped, to 0.9 x 0.4 cm; operculum conical to beaked. **Flowers** white, January to April. **Fruit** stalked, cup-shaped to barrel-shaped, to 0.9 x 0.6 cm.

DISTRIBUTION Widespread on coastal plains and ranges northwards from Jervis Bay in New South Wales to the area between Rockhampton and Mackay in Queensland, then with a substantial gap to the northern occurrences in the ranges from west of Paluma to the southern part of the Atherton Tableland; also two small, disjunct patches, east of Clermont and near Eungella Dam.

NOTES An eastern box recognised by smooth upper trunk and limbs. Leaves are large, glossy, green. It is a wet-country relative of the white box (*E. albens*).

143

Eucalyptus viridis

GREEN MALLEE

DESCRIPTION Mallee. **Bark** rough and persistent on the lower part of the stems, dark grey, ribbony above, finally smooth, white, grey and pinkish grey; may be mostly smooth in smaller mallees, with basal accumulations of shed bark ribbons. **Juvenile leaves** very shortly stalked, linear to narrow-lanceolate, to 12 x 0.6 cm, green. **Adult leaves** shortly stalked, linear, to 15 x 0.8 cm, concolorous, glossy, dark green, with numerous oil glands. **Buds** in axils of leaves or sometimes clustered at ends of branchlets, in 7s, stalked, ovoid or broadly spindle-shaped, to 0.7 x 0.3 cm; operculum conical. **Flowers** white, November to January. **Fruit** stalked, cup-shaped to truncate-globose, to 0.5 x 0.4 cm.

DISTRIBUTION Geographically widespread but very scattered occurrences, from north-west of Taroom in Queensland, through the central and north-western slopes and plains of New South Wales, in parts of northern and western Victoria, and in the Flinders Ranges of South Australia.

NOTES An attractive mallee box with its narrow, green leaves.

Eucalyptus argophloia

BURNCLUITH GUM,
QUEENSLAND WESTERN WHITE GUM

DESCRIPTION Medium-sized to tall tree. **Bark** smooth, grey and reddish grey over yellow, weathering to white, powdery. **Juvenile leaves** stalked, linear to narrow-lanceolate, to 9 x 1.4 cm, greyish green. **Adult leaves** stalked, narrow-lanceolate, to 13 x 1.3 cm, concolorous, glossy, green. **Buds** in axils of leaves, in 7s, stalked, ovoid to globular, to 0.4 x 0.4 cm; operculum hemispherical to conical. **Flowers** white, May to August. **Fruit** stalked, hemispherical to cup-shaped, to 0.5 x 0.7 cm; valves 5 or 6.

DISTRIBUTION Of very restricted natural distribution in south-eastern Queensland in the region north-east of Chinchilla, including Burncluith, Pelican and Burra Burri; on heavy clay soils.

NOTES A beautiful tall straight white-barked tree, becoming rare in its natural occurrence which is prime agricultural land. Fruit are small with up to 6 valves.

Eucalyptus bosistoana

COAST GREY BOX

DESCRIPTION Medium-sized to tall tree. **Bark** variable, usually rough and persistent on the lower part or most of the trunk, thin, short-fibred, light grey; shedding above in ribbons to leave a smooth, light grey to whitish surface. **Juvenile leaves** stalked, ovate to round, to 10 x 9.5 cm, pale green. **Adult leaves** stalked, lanceolate to narrow-lanceolate, to 20 x 1.8 cm, concolorous, dull or slightly glossy, green. **Buds** stalked, in simple clusters in axils of leaves as well as compound clusters in axils and at ends of branchlets, in 7s, stalked, broadly club-shaped, to 0.9 x 0.4 cm; operculum conical or hemispherical. **Flowers** white, November to February. **Fruit** stalked, cup-shaped, hemispherical or barrel-shaped, to 0.7 x 0.7 cm; valves 5 or 6.

DISTRIBUTION Occurs on coastal plains and nearby ranges from near Sydney, New South Wales, southwards to the eastern Gippsland region of Victoria.

NOTES An eastern box species, the bark is difficult to categorise, as it is often thin and only partly retained. Easily distinguished in its area of distribution by the 5 or 6 valves of the fruit.

Eucalyptus froggattii

KAMAROOKA MALLEE

DESCRIPTION Small tree or mallee. **Bark** rough and persistent over the trunk, fibrous, with shallow longitudinal fissures, grey or greyish brown, shedding above in strips to leave a smooth, brownish or yellowish surface; mallees mostly smooth-barked. **Juvenile leaves** stalked, lanceolate to broad-lanceolate, to 10 x 4.5 cm, glossy, light green. **Adult leaves** stalked, lanceolate, to 12 x 1.5 cm, concolorous, glossy, light green, with numerous oil glands. **Bud clusters** in axils of leaves and at ends of branchlets, in 7s to 11s. **Buds** stalked, club-shaped or ovoid in outline, square in cross-section, to 1 x 0.5 cm; operculum pyramidal. **Flowers** white, January to April. **Fruit** stalked, cup-shaped or cylindrical in outline, square in cross-section, to 1 x 0.8 cm.

DISTRIBUTION Of restricted distribution in Victoria, in the Kamarooka district north of Bendigo and to the north-west of Charlton.

NOTES A mallee box easy to recognise by the large 'square' buds and fruits, although it may grade into other species. A character in the species unique in the mallee boxes is the great density of oil glands in the leaves, which are so numerous as to obscure the venation.

Eucalyptus ochrophloia

YAPUNYAH

DESCRIPTION Small to medium-sized tree. **Bark** rough on lower part or all of trunk and sometimes on the larger limbs, dark grey; smooth, grey, whitish or coppery above. **Juvenile leaves** shortly stalked, linear to narrow-lanceolate, to 16 x 1.4 cm, dull, blue-green. **Adult leaves** stalked, lanceolate, to 19 x 2.5 cm, concolorous, glossy, green, with numerous oil glands. **Buds** mostly in compound clusters towards the ends of the branchlets and in axils of leaves, but also some simple clusters in leaf axils, in 7s, stalked, broadly spindle-shaped or elongated and often curved, ribbed, to 1.5 x 0.5 cm; operculum conical. **Flowers** white, June to November. **Fruit** stalked, cylindrical to barrel-shaped, to 1.8 x 0.7 cm.

DISTRIBUTION Restricted to major watercourses in south-western Queensland from Cunnamulla to Noccundra, to north of Adavale. Common along Paroo River, Kyabra Creek, Bulloo River, Wilson River, Warrego River and tributaries; also north-western New South Wales along the Paroo River.

NOTES A major honey-producing species. The upper smooth bark is very attractive when, in season, it is a bright coppery colour.

Eucalyptus decorticans

GUM-TOPPED IRONBARK

DESCRIPTION Medium-sized tree. **Ironbark** on trunk and larger limbs, grey to dark grey; conspicuously smooth above, white to coppery. **Juvenile leaves** stalked, lanceolate, to 11 x 1.7 cm, green. **Adult leaves** stalked, narrow-lanceolate to lanceolate, to 12 x 1.5 cm, concolorous, dull, light green. **Bud clusters** in axils of leaves and at ends of branchlets, in 7s and 9s. **Buds** stalked, club-shaped to broadly spindle-shaped, to 0.7 x 0.4 cm; operculum conical. **Flowers** white, December to January. **Fruit** stalked, cup-shaped to truncate-ovoid, to 0.7 x 0.7 cm.

DISTRIBUTION Widespread in south-eastern Queensland, from west of the Drummond Range, east to Monto, south-east to Kingaroy, and west of Millmerran, with an outlier in the Callide Range north of Biloela; always in sandstone hills.

NOTES One of many Queensland ironbark species, it is usually recognised by the site and by its ironbark and smooth upper branches.

149

3 cm, concolorous, dull or slightly glossy, green to bluish green or greyish green. **Buds** in branched clusters in axils of leaves and at ends of branchlets, with some unbranched in leaf axils, in 7s to 11s, stalked, broadly spindle-shaped, to 2 x 0.5 cm; operculum conical to horn-shaped. **Flowers** white, April to July. **Fruit** stalked, funnel-shaped or pear-shaped, to 1.2 x 1 cm.

DISTRIBUTION From north-west of Rockhampton in Queensland, southwards, inland as far as Dalby and to west of Mundubbera; also coastal plains and hills in New South Wales to around Bodalla on the south coast.

NOTES The form from north of Rockhampton is notable for very thick stalks on the buds and fruit. The closely related *E. fibrosa* subsp. *nubila* differs in having glaucous, bluish leaves at all stages, and glaucous buds and fruit.

Eucalyptus fibrosa

BROAD-LEAVED RED IRONBARK, RED IRONBARK, BROAD-LEAVED IRONBARK

DESCRIPTION Medium-sized to tall tree. **Ironbark** persistent on trunk and branches, deeply and coarsely furrowed, hard or flaky, grey to black. **Juvenile leaves** stalked, ovate to round, to 21 x 12 cm, green. **Adult leaves** stalked, lanceolate to broad-lanceolate, to 18 x

Eucalyptus crebra

NARROW-LEAVED RED IRONBARK, IRONBARK, NARROW-LEAVED IRONBARK

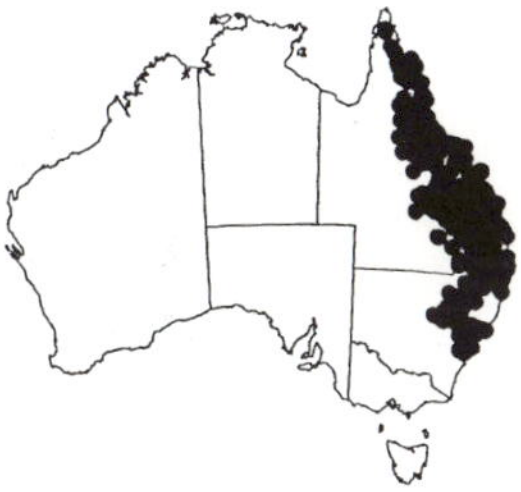

DESCRIPTION Small to medium-sized or occasionally tall tree. **Ironbark** persistent to the small branches, deeply and coarsely furrowed, hard, grey to dark grey. **Juvenile leaves** stalked, narrow-lanceolate, to 12.5 x 2 cm, green or bluish green. **Adult leaves** stalked, narrow-lanceolate to lanceolate, to 15 x 1.5 cm, concolorous, dull, green to bluish green. **Buds** in branched clusters in axils of leaves and at ends of branchlets, in 7s to 11s, stalked, club-shaped or diamond-shaped, to 0.8 x 0.4 cm; operculum conical to hemispherical. **Flowers** white, May to November. **Fruit** stalked, hemispherical or cup-shaped, to 0.7 x 0.6 cm.

DISTRIBUTION The most widespread of all ironbark species, occurring along the central coast and north-western slopes and plains of New South Wales, and in eastern Queensland as far north as Cooktown, inland to Einasleigh, Pentland, Alpha, Mitchell and Goondiwindi.

NOTES Probably one of the most difficult of all ironbarks to categorise because of its many forms that differ in leaf size and colour, and bud and fruit shape. It will probably be divided into several subspecies.

Eucalyptus melanophloia

SILVER-LEAVED IRONBARK

DESCRIPTION Small to medium-sized tree, rarely a mallee. **Ironbark** persistent to small limbs, grey, dark grey or black. **Juvenile leaves** stalkless, opposite, ovate to round, to 10 x 10 cm, bluish, glaucous. **Adult leaves** very shortly stalked, opposite, ovate to broad-lanceolate, to 9 x 3.5 cm, concolorous, bluish to glaucous. **Buds** in branched clusters in axils of leaves and at ends of branchlets, in 7s, stalked, usually glaucous, double-conic, to 0.8 x 0.4 cm; operculum conical. **Flowers** white, September to February. **Fruit** stalked, cupular or hemispherical to truncate-globose, occasionally urn-shaped, usually glaucous, to 0.6 x 0.6 cm.

DISTRIBUTION Occurring extensively in Queensland from Brisbane to west of Charleville and northwards to north of Chillagoe and west of Croydon, also as a small mallee in an isolated patch south of Mount Isa; widely distributed from the western slopes of the northern tablelands of New South Wales to north of Bourke.

NOTES With its round glaucous leaves, the easiest to recognise of all the ironbarks. The related *E. shirleyi* has larger leaves, buds and fruits.

152

Eucalyptus baueriana

ROUND LEAF BOX

DESCRIPTION Small to medium-sized tree. **Bark** rough and persistent on the trunk and branches, fibrous, flaky, light grey, with paler, almost white patches. **Juvenile leaves** stalked, round to broadly ovate, undulate, to 10 x 7.5 cm, light green, thin. Trees usually mature in juvenile leaf phase. **Adult leaves** stalked, ovate to broad-lanceolate, undulate, to 10 x 6 cm, concolorous, green, thin. **Buds** in branched clusters in axils of leaves and at ends of branchlets, in 7s, stalked, club-shaped or diamond-shaped, to 0.6 x 0.3 cm; operculum conical. **Flowers** white, November to December. **Fruit** stalked, funnel-shaped, to 0.7 x 0.6 cm.

DISTRIBUTION From coastal eastern Victoria (with a few inland occurrences) from around Bairnsdale, extending eastwards into New South Wales to north of Sydney.

NOTES This is the coastal, wetter-country species of the 'red boxes'. It differs from *E. polyanthemos* by the habitat and glossy green leaves.

153

Eucalyptus paniculata

GREY IRONBARK

DESCRIPTION Medium-sized to tall tree. **Ironbark** persistent to the small branches, furrowed, hard or corky, light grey. **Juvenile leaves** stalked, ovate, to 16 x 6 cm, green. **Adult leaves** stalked, lanceolate, to 15 x 2.3 cm, discolorous, slightly glossy, green. **Buds** in branched clusters at ends of branchlets, in 7s, stalked, ovoid to diamond-shaped, to 1.1 x 0.5 cm; operculum conical, often narrower than base at join. **Flowers** white, May to January. **Fruit** stalked, hemispherical, funnel-shaped or barrel-shaped, to 0.9 x 0.7 cm.

DISTRIBUTION Coastal New South Wales from around Bega in the south extending northwards to the Coffs Harbour area.

NOTES One of two New South Wales ironbark species with discolorous adult leaves, the other being *E. dorsiventralis,* which has larger, more strongly ribbed buds and fruit. This and related species have a deciduous ring of tissue usually retained on the rim of the fruit.

154

Eucalyptus sideroxylon

RED IRONBARK, MUGGA IRONBARK, MUGGA

DESCRIPTION Small to medium-sized or occasionally tall tree. **Ironbark** persistent on trunk and larger branches, hard and deeply furrowed, dark grey to black; upper limbs smooth, whitish. **Juvenile leaves** stalked, narrow-lanceolate, to 15 x 2 cm, sub-glaucous or dull green. **Adult leaves** stalked, lanceolate, to 14 x 1.8 cm, concolorous, dull, green, glaucous or slate grey. **Buds** pendulous in axils of leaves, in 7s, with long, slender stalks, club-shaped to diamond-shaped, to 1.2 x 0.5 cm; operculum conical or beaked. **Flowers** white, pink, red or pale yellow, May to November. **Fruit** on long, slender stalks, cup-shaped or barrel-shaped or truncate-globose, to 1 x 0.9 cm; rim with black deciduous ring.

DISTRIBUTION South-eastern Queensland from Texas and Wallangarra to north of Tara, with scattered occurrences further north; also widespread on the western slopes and plains of New South Wales, extending just into Victoria near Chiltern; also west of Sydney towards the Blue Mountains.

NOTES Very conspicuous with its usually black ironbark. Widely planted for its coloured flowers.

Eucalyptus melliodora

YELLOW BOX, HONEY BOX, YELLOW IRONBARK

DESCRIPTION Medium-sized to occasionally tall tree. **Bark** very variable from mostly smooth with an irregular short stocking, to covering most of the trunk, fibrous, dense or loosely held, grey, yellow- or red-brown, occasionally very coarse, thick, dark brown to black; shedding from the upper trunk and branches to leave a smooth, whitish or yellowish surface. **Juvenile leaves** stalked, elliptical, to 11 x 5 cm, grey-green. **Adult leaves** stalked, narrow-lanceolate to lanceolate, to 14 x 1.8 cm, concolorous, dull, light green or slate grey. **Intramarginal vein** in both juvenile and adult leaves is conspicuously remote from the leaf edge. **Buds** usually in axils of leaves, often clustered towards the tips of branchlets, in 7s, on slender stalks, club-shaped to ovoid, to 0.8 x 0.4 cm; operculum conical or beaked. **Flowers** white or occasionally pink, September to February. **Fruit** stalked, hemispherical or truncate-globose, to 0.7 x 0.7 cm; rim with black deciduous ring.

DISTRIBUTION Widely distributed on plains and tablelands from western Victoria through New South Wales to south-central Queensland.

NOTES A very important honey-producing species.

156

Eucalyptus leucoxylon

BLUE GUM, YELLOW GUM, WHITE IRONBARK

DESCRIPTION Small to medium-sized tree. **Bark** usually rough and persistent for 1–2 m, fibrous, shedding above to a smooth, white, yellow or bluish grey surface. **Juvenile leaves** stalkless for many pairs (joined around the stem (connate) in some Victorian localities), to 9 x 6 cm, dull green. **Adult leaves** stalked, lanceolate to broad-lanceolate, to 13 x 2.5 cm, concolorous, dull, green. **Buds** in axils of leaves, in 3s, on long, slender stalks, club-shaped or diamond-shaped, sometimes warty, to 1.5 x 0.8 cm; operculum conical or beaked. **Flowers** white, pink or red, May to September. **Fruit** stalked, barrel-shaped or truncate-globose, to 1.2 x 1.2 cm; rim with black deciduous ring (staminophore).

DISTRIBUTION Widely distributed on plains and nearby ranges of coastal South Australia (blue gum), extending into the western half of Victoria (yellow gum), with isolated small patches just into New South Wales.

NOTES This species has been divided into numerous varieties and subspecies. A spectacular red-flowered form of uncertain provenance is widely planted as an ornamental, flowering profusely in winter.

Eucalyptus raveretiana

BLACK IRONBOX

DESCRIPTION Small to medium-sized tree. **Bark** rough, longitudinally fissured, grey over brown on trunk and base of larger limbs; smooth, bluish grey above. **Juvenile leaves** stalked, ovate, to 10 x 3 cm, green. **Adult leaves** stalked, narrow-lanceolate to lanceolate or ovate, to 15 x 3 cm, strongly discolorous, dull, green to dark green or bluish green. **Buds** in clusters in axils of leaves and at ends of branchlets, in 7s, with slender stalks, elongated, to 0.4 x 0.15 cm; operculum conical. **Flowers** white, December to February. **Fruit** distinctly stalked, shallowly hemispherical, to 0.2 x 0.2 cm; valves prominently exserted.

DISTRIBUTION Scattered and disjunct in coastal and sub-coastal central Queensland, from Dipperu National Park south-west of Mackay, northwards and north-westwards almost to Charters Towers, Bowen and Ayr, also Rockhampton and westwards and near the Mackenzie River north of Duaringa. Always along creeks and rivers.

NOTES One of three northern species of scattered distribution and notable for the very small buds and fruits.

Eucalyptus brachyandra

TROPICAL RED BOX

DESCRIPTION Small straggly, often gnarled tree. **Bark** rough on trunk and base of larger branches, stringy, wavy, grey over yellowish brown; smooth, whitish or grey above. **Juvenile leaves** stalked, opposite for many pairs, ovate, to 10 x 7 cm, light green. **Adult leaves** stalked, elliptical to round, to 8 x 4 cm, discolorous, dull, light green, with dense reticulation. **Buds** in clusters in axils of leaves and at ends of branchlets, in 7s, distinctly stalked, pear-shaped to 0.3 x 0.2 cm; operculum shallowly conical to hemispherical. **Flowers** white, September to October. **Fruit** stalked, cup-shaped, to 0.25 x 0.2 cm.

DISTRIBUTION Scattered occurrence in the north-western region of the Northern Territory (e.g. Victoria River area, Jasper Gorge, Port Keats, Tabletop Range), and in the Kimberley region of Western Australia (e.g. Mt Elizabeth Station, Berkeley River, Hidden Valley near Kununurra, Prince Regent River, Sunday Island, Bungle Bungles). Always on very rocky sites, not uncommonly jutting out of small crevices in sheer cliff faces.

NOTES One of three northern species of scattered distribution and notable for the very small buds and fruits.

Eucalyptus howittiana

HOWITT'S BOX

DESCRIPTION Medium-sized tree. **Bark** rough, hard, scaly to smaller limbs, grey to whitish grey over rich brown. **Juvenile leaves** stalked, ovate to broad-lanceolate, to 10 x 6 cm, glossy, green. **Adult leaves** stalked, broad-lanceolate, to 10 x 2.5 cm, discolorous, very glossy, green, with dense reticulation. **Buds** in crowded clusters towards ends of branchlets, in 7s, stalkless, spindle-shaped, to 0.5 x 0.1 cm; operculum conical to beaked. **Flowers** creamy white, January to May. **Fruit** stalkless, cup-shaped to shortly barrel-shaped or truncate-globose, to 0.25 x 0.25 cm.

DISTRIBUTION Restricted to two small areas of north-central Queensland: north-west of Wairuna to south of Greenvale, including Valley of Lagoons; and from 'Mt Cooper' to 'Scartwater', south-east of Charters Towers. Usually on breakaways or rocky outcrops.

NOTES One of three northern species of scattered distribution and notable for the very small buds and fruits.

Eucalyptus microcorys

TALLOWWOOD

DESCRIPTION Medium-sized to tall tree. **Bark** rough and persistent to the small branches, moderately long-fibred, spongy, with longitudinal fissures, reddish brown over inner yellow-brown; underbark flaky with minute horizontal cracks, outer bark with small, circular pores. **Juvenile leaves** stalked, elliptical to ovate, to 11 x 5 cm, green, thin. **Adult leaves** stalked, lanceolate to 12 x 2.5 cm, discolorous, glossy, dark green, thin. **Buds** in axils of leaves, often clustered towards ends of branchlets, in 7s and 9s, on long, tapering stalks, club-shaped, to 0.6 x 0.3 cm; operculum hemispherical with cross-sutures. **Flowers** creamy white, July to November. **Fruit** on tapering stalks, funnel-shaped, to 0.9 x 0.6 cm.

DISTRIBUTION Coast and nearby ranges from around Newcastle in New South Wales to Maryborough in Queensland.

NOTES A fine forest tree with valuable timber, it is not closely related to any other species. It is recognised by the reddish rough bark, discolorous leaves and funnel-shaped fruits.

Eucalyptus curtisii

PLUNKETT MALLEE

DESCRIPTION Slender to stout-stemmed mallee. **Bark** smooth, greenish, light grey, white or coppery. **Juvenile leaves** stalked, lanceolate, to 10 x 1.5 cm, green. **Adult leaves** stalked, lanceolate, to 13 x 2.5 cm, strongly discolorous, dark green. **Buds** in clusters towards ends of branchlets, in 7s, stalked, ovoid to club-shaped, to 0.8 x 0.7 cm; operculum hemispherical to shallowly conical. **Flowers** white, September to December. **Fruit** stalked, bell-shaped or cup-shaped, to 1 x 0.8 cm. Seed long and narrow.

DISTRIBUTION Wide but very scattered distribution in south-eastern Queensland from Robinson Gorge and Isla Gorge National Parks through Barakula State Forest to the Plunkett region south of Brisbane.

NOTES Related to no other eucalypt, it is notable for the discolorous leaves, terminal bud clusters, narrow seed, and fruit which do not produce valves but shed a star-shaped disc from the top. It grows well as an ornamental, even in quite cold regions (e.g. Canberra) well away from its natural area.

162

Eucalyptus tenuipes

NARROW-LEAVED WHITE MAHOGANY

DESCRIPTION Small to medium-sized tree. **Bark** rough, fibrous or stringy to the small branches, brown or grey-brown over reddish brown and yellowish. **Juvenile leaves** opposite for many pairs, shortly stalked or without stalks, linear, to 12 x 0.8 cm. **Adult leaves** shortly stalked, lanceolate, to 10 x 1 cm, discolorous, dull, green, with dense reticulation. **Buds** in axils of leaves, in 7s or more, on long, slender stalks, ovoid to club-shaped, to 0.4 x 0.4 cm; operculum hemispherical. **Flowers** white, March to June. **Fruit** on long, slender stalks, cup-shaped to truncate-globose or urn-shaped, to 0.5 x 0.5 cm.

DISTRIBUTION Abundant in Queensland from south-east of Springsure to Chinchilla but occurring in a few isolated stands to the north, e.g. Fork Lagoons and Junee Tableland, with an outlier at 'Allandale' north-east of Pentland.

NOTES Related to no other eucalypt, despite the common name, it is recognised by the fibrous bark, discolorous leaves and the long, slender stalks of the buds and fruit.

Eucalyptus cloeziana

GYMPIE MESSMATE

DESCRIPTION Small to tall tree. **Bark** rough on most or whole of trunk and larger branches, soft, flaky, tessellated, light brown or yellow-brown; smooth, greyish white or yellowish above. **Juvenile leaves** stalked, elliptical to ovate, to 10 x 5 cm, greyish green. **Adult leaves** stalked, lanceolate or curved, to 13 x 3 cm, discolorous, slightly glossy, green, with fine, dense reticulation. **Buds** in leafless compound clusters in axils of leaves although often maturing towards ends of branchlets, in 7s, stalked, ovoid to club-shaped, to 0.7 x 0.4 cm; operculum hemispherical to obtusely conical. **Flowers** white, November to February. **Fruit** stalked, hemispherical, to 1 x 1.2 cm, with broad disc.

DISTRIBUTION Widespread but scattered in Queensland particularly from east of Tambo to Mundubbera and Gympie, and from west of Townsville to north-west of Cooktown with isolated occurrences in between.

NOTES Reaches its best development in the Gympie region, where some magnificent forest trees attain heights close to 60 m. In other places it can be a poorly formed tree of only 10–15 m. The flaky bark can be confused with that of yellow bloodwoods, to which it is quite unrelated.

Eucalyptus acmenoides

WHITE MAHOGANY

DESCRIPTION Medium-sized to tall tree. **Bark** rough, fibrous, not as coarse as a stringybark, often held in somewhat flattish strips, whitish grey to grey-brown. **Juvenile leaves** stalkless, stem-clasping, opposite for many pairs, ovate, to 13 x 5 cm. **Adult leaves** stalked, lanceolate, to 12 x 2.5 cm, concolorous or discolorous, slightly glossy, green, thin, with dense reticulation. **Buds** mostly in axils of leaves, but also clustered towards the ends of branchlets, in 7s or more, stalked, ovoid to broadly spindle-shaped, to 0.7 x 0.3 cm; operculum conical or beaked. **Flowers** white, October to December. **Fruit** stalked, barrel-shaped, urn-shaped or hemispherical, to 0.8 x 0.7 cm.

DISTRIBUTION Largely coastal and sub-coastal including ranges from north of Sydney to north of Rockhampton and then from Townsville to Cooktown, with isolated, disjunct occurrences in between.

NOTES One of several white mahoganies (unrelated to the red mahoganies), it is the most important commercially. The rough bark and thin, discolorous leaves indicate its relatively primitive status in the monocalypts (pp. 185–221). The dense reticulation of the leaves, also a primitive character, distinguishes the white mahoganies from other more advanced monocalypts.

165

Eucalyptus macrorrhyncha

RED STRINGYBARK

DESCRIPTION Small to medium-sized tree. **Bark** rough, thick, fibrous and stringy, dark brown. **Juvenile leaves** stalkless or with short stalks, ovate, to 12 x 5 cm, undulate; early leaves covered with fine hairs. **Adult leaves** stalked, lanceolate, to 15 x 2.5 cm, concolorous, slightly glossy, green. **Buds** in axils of leaves, in 7s or more, stalked, diamond-shaped, to 0.9 x 0.5 cm; operculum beaked. **Flowers** white, January to April. **Fruit** stalked, globular (including the disc), to 1 x 1.2 cm; disc broad, ascending.

DISTRIBUTION Occurring on ranges and tablelands of New South Wales and Victoria, with a small, disjunct population south-west of Clare in South Australia.

NOTES One of a large number of stringybarks that occur mostly in New South Wales. It is distinguished by the beaked buds and +/– globular fruit with the massive disc.

Eucalyptus agglomerata

BLUE-LEAVED STRINGYBARK

DESCRIPTION Medium-sized to tall tree. **Bark** rough, thick, fibrous and stringy, usually conspicuously furrowed, grey over red-brown. **Juvenile leaves** shortly stalked, ovate, to 15 x 6 cm, green; early leaves covered with fine hairs. **Adult leaves** stalked, lanceolate to broad-lanceolate, often curved and oblique, to 14 x 3 cm, concolorous, slightly glossy, usually bluish green. **Buds** in axils of leaves, in 11s or more, stalkless, spindle-shaped, to 0.8 x 0.3 cm; operculum conical. **Flowers** white, March to August. **Fruit** stalkless, crowded, broader than long or hemispherical, sides flattened by crowding, to 0.6 x 1 cm.

DISTRIBUTION Occurring on the central tablelands and central and southern coast of New South Wales, extending into the far eastern corner of Victoria.

NOTES One of the seven or eight stringybarks in Victoria. It is recognised by the bluish tone to the leaves, seen particularly in the whole crown, and the tightly clustered, flattened fruits.

Eucalyptus baxteri

BROWN STRINGYBARK

DESCRIPTION Varying greatly in form from a shrub or mallee to a tall tree. **Bark** rough, thick, fibrous and stringy, longitudinally furrowed, grey or brownish. **Juvenile leaves** stalkless, ovate, often with a distinct point at the leaf tip, to 13 x 8 cm, undulate, green; early leaves and stems covered with fine hairs. **Adult leaves** stalked, broadly curved to broad-lanceolate, to 13 x 3 cm, concolorous, slightly glossy, green, relatively thick. **Buds** in axils of leaves, in 7s or more, stalkless or shortly stalked, oblong to club-shaped, warty, to 0.8 x 0.5 cm; operculum hemispherical or conical. **Flowers** white, February to March. **Fruit** stalkless or shortly stalked, hemispherical or sometimes truncate-globose, to 1.1 x 1.6 cm; disc broad, level or ascending.

DISTRIBUTION Widespread in hilly country from the Mount Lofty Ranges of South Australia to the far south-east of New South Wales.

NOTES A form from sandy 'desert' country of South Australia and Victoria is now recognised as a distinct species, *E. arenacea*.

Eucalyptus serraensis

GRAMPIANS GUM

DESCRIPTION Mallee or rarely a small tree. **Bark** flaky with upper trunk and branches smooth. **Juvenile leaves** stalked, at first with hairs, broadly ovate to round, to 6 x 4.5 cm, glossy, green. **Adult leaves** stalked, ovate to broad-lanceolate, to 11 x 5 cm, glossy, green, thick. **Buds** in axils of leaves, in 3s or 7s, stalkless, club-shaped, warty, to 1 x 1 cm; operculum hemispherical. **Flowers** white, December to April. **Fruit** stalkless, hemispherical, to 1.2 x 1.8 cm; disc broad, ascending; valves exserted.

DISTRIBUTION Peaks and rocky outcrops of the Serra, Wonderland and possibly Mt William Range of the Grampians in Victoria.

NOTES When it was found that the original specimen of *E. alpina* was the same as *E. baxteri,* a 'new' name was created for this well-known (especially in cultivation) species of the Grampians.

Eucayptus olsenii

WOILA GUM

DESCRIPTION Small tree. **Bark** rough at the base, fibrous or flaking, greyish; upper trunk and limbs smooth, grey, white or yellowish. **Juvenile leaves** stalkless and opposite at first, becoming alternate and shortly stalked, broad-lanceolate, to 10 x 3.5 cm, glossy, green; early leaves and stems covered with fine hairs. **Adult leaves** stalked, lanceolate or curved and oblique, to 12 x 2 cm, concolorous, glossy, green, with numerous oil glands. **Buds** in axils of leaves, in 7s, stalkless or shortly stalked, club-shaped, with prominent ribs, to 1.5 x 0.8 cm; operculum conical. **Flowers** white, September to December. **Fruit** stalkless, barrel-shaped, ribbed, to 2 x 1.8 cm.

DISTRIBUTION Very restricted in mountains north-east of Cooma in New South Wales, in the area bounded approximately by the peaks Dampier, Mother Woila, The Scout and Tabletop.

NOTES A curious species without certain affinities. It is found only on very rough rocky terrain, but grows well as an ornamental, e.g. in Canberra.

Eucalyptus pilularis

BLACKBUTT

DESCRIPTION Medium-sized to very tall tree. **Bark** rough over part or most of the trunk, finely fibrous, spongy, brown or grey over red-brown; bark above smooth, creamy white, often with insect 'scribbles'. **Juvenile leaves** stalkless, stem-clasping, opposite for many pairs, broad-lanceolate, to 17 x 3.5 cm, strongly discolorous, green above, often purple below; young stems squarish in cross-section. **Adult leaves** stalked, lanceolate or curved and oblique, to 16 x 3 cm, glossy, green to dark green, with numerous oil glands. **Buds** in axils of leaves, in 7s to 15s, stalked, broadly spindle-shaped, to 1.1 x 0.5 cm; operculum conical. **Flowers** white, September to March. **Fruit** stalked, hemispherical to truncate-globose, to 1.1 x 1.1 cm.

DISTRIBUTION Coast and nearby ranges from southern New South Wales to Fraser Island in Queensland.

NOTES A very important commercial timber species. In size and numbers, it dominates many of the east coast forests and can be recognised by the rough, fibrous bark on most of the trunk with smooth, white bark above.

Eucalyptus regnans

MOUNTAIN ASH, SWAMP GUM, STRINGY GUM

DESCRIPTION Tall to very tall tree. **Bark** rough over lower part of trunk, thin, fibrous, brown, then smooth, cream, grey or greenish, often with long ribbons of shedding dead bark. **Juvenile leaves** stalked, ovate to broad-lanceolate, to 17 x 8 cm, glossy, green. **Adult leaves** stalked, lanceolate or curved and oblique, to 14 x 2.7 cm, concolorous, green. **Inflorescences** mostly in pairs in axils of leaves, in 7s to 15s or more. **Buds** stalked, club-shaped, to 0.7 x 0.4 cm; operculum conical. **Flowers** white, December to May. **Fruit** stalked, obconical or pear-shaped, to 0.9 x 0.7 cm.

DISTRIBUTION Occurring in forests of the Otway Ranges south-west of Melbourne, mountains in eastern Victoria and lower terrain in several regions of Tasmania.

NOTES The tallest hardwood tree in the world, occasionally to 100 m in height. It and the following species, *E. fastigata,* are two of only four species in the genus that regularly have paired inflorescences. Distinguished from *E. fastigata* by having less rough bark and by the fruit with slightly sunken disc.

Eucalyptus fastigata

BROWN BARREL, CUT TAIL

DESCRIPTION Medium-sized to very tall tree. **Bark** rough on trunk and larger limbs, thick, fibrous, often coarsely furrowed and somewhat stringy, shedding above in long ribbons which often remain hanging in the crown, to leave a smooth, whitish surface. **Juvenile leaves** stalked, ovate to broad-lanceolate, oblique, to 17 x 5 cm, glossy, green. **Adult leaves** stalked, lanceolate or curved and oblique, to 15 x 2.7 cm, concolorous, green. **Inflorescences** mostly paired in axils of leaves, in 11s to 15s or more. **Buds** stalked, club-shaped, to 0.6 x 0.3 cm; operculum conical or slightly beaked. **Flowers** white, December to February. **Fruit** stalked, obconical or pear-shaped, to 0.8 x 0.7 cm.

DISTRIBUTION Occurring on ranges, tablelands and escarpments of New South Wales and adjacent parts of far eastern Victoria.

NOTES Distinguished from *E. regnans* by the rough bark covering the whole trunk and larger limbs, and by the slightly domed disc of the fruit.

173

Eucalyptus obliqua

MESSMATE STRINGYBARK, MESSMATE, STRINGYBARK

DESCRIPTION Varying from a small to very tall tree. **Bark** rough, thick, fibrous, variable in appearance, sometimes stringy and at other times compact and irregularly fissured; at high altitudes in Tasmania the smaller branches may be smooth, grey-green. **Juvenile leaves** stalked, broadly ovate, oblique, to 19 x 7.5 cm, glossy, green. **Adult leaves** stalked, broad-lanceolate or curved, to 15 x 3.3 cm, concolorous, glossy, green. **Buds** in axils of leaves, in 11s or more, stalked, club-shaped, to 0.7 x 0.4 cm; operculum hemispherical with a small point. **Flowers** white, December to March. **Fruit** stalked, barrel-shaped or sometimes truncate-globose, to 1.1 x 0.9 cm.

DISTRIBUTION Widespread; the Mount Lofty Ranges, Kangaroo Island and south-east of South Australia; the coast and lower ranges of Victoria; most of Tasmania; the south coast, coastal ranges and northern tablelands of New South Wales, just extending into Queensland.

NOTES The first eucalypt to be named, from a specimen collected on Bruny Island, Tasmania, in 1777. The name *obliqua* refers to the asymmetrical leaves, a characteristic that is, however, widespread in the genus.

Eucalyptus stricta

BLUE MOUNTAINS MALLEE ASH

DESCRIPTION Mallee. **Bark** smooth, whitish, greyish or greenish, usually with an accumulation of shed bark strips at the base. **Juvenile leaves** stalked, broad-lanceolate, often oblique, to 10 x 4 cm, glossy, green. **Adult leaves** stalked, lanceolate to narrow-lanceolate, to 10 x 1 cm, concolorous, glossy, green, usually held vertically. **Buds** in axils of leaves, in 7s, stalked, club-shaped, to 0.7 x 0.3 cm; operculum hemispherical, usually with a small point. **Flowers** white, December to April. **Fruit** barrel-shaped to slightly urn-shaped, to 0.9 x 0.8 cm.

DISTRIBUTION Central Tablelands and coast of New South Wales south to the Bega area, particularly on sandstone plateaus.

NOTES One of the green-leaved mallee ashes. It occupies a position between the narrower-leaved *E. apiculata* and broader-leaved *E. obstans*.

Eucalyptus stellulata

BLACK SALLY

DESCRIPTION Small to medium-sized tree. **Bark** rough on lower half of the trunk, compact, dark grey to grey-black; smooth, oily, yellow-green to olive-green above. **Juvenile leaves** stalkless and opposite for many pairs, round to ovate, to 9 x 6 cm, dull, green; later leaves with short stalks and becoming alternate. **Adult leaves** stalked, elliptical to broad-lanceolate, with a small pointed, hooked tip, to 8 x 2.5 cm, concolorous, glossy, green, with 3 veins. **Buds** in axils of leaves, in > 7s, stalkless or shortly stalked, spindle-shaped, to 0.6 x 0.2 cm; operculum conical. **Flowers** white, April to October. **Fruit** virtually stalkless, crowded, cup-shaped to truncate-globose, to 0.5 x 0.5 cm.

DISTRIBUTION Widespread in high tablelands and sub-alpine situations of New South Wales from the far north of the State extending southwards into eastern Victoria; favouring swamps and flats beside streams, but it can also be found on rocky sites on high mountains.

NOTES Belonging to a small group of species that includes *E. moorei*, it is unrelated to the white sallies (snow gums). The tight, star-shaped bundles of buds are characteristic.

Eucalyptus elata

RIVER PEPPERMINT

DESCRIPTION Medium-sized to tall tree. **Bark** rough on lower part of trunk, compact with narrow longitudinal fissures, dark grey, shedding above in long ribbons, often remaining partly attached in the crown, to leave a smooth, grey, cream or whitish surface. **Juvenile leaves** stalkless, stem-clasping, opposite for many pairs, broad-lanceolate to lanceolate, to 12 x 2.8 cm, green. **Adult leaves** stalked, narrow-lanceolate, to 15 x 1.5 cm, concolorous, glossy green, thin, with a strong peppermint smell when crushed. **Buds** in axils of leaves, often more than 20 in each group, on long, slender stalks, club-shaped, to 0.5 x 0.25 cm; operculum hemispherical or conical. **Flowers** white, September to November. **Fruit** stalked, usually crowded, truncate-globose, to 0.6 x 0.6 cm.

DISTRIBUTION Occurs on the central tablelands and southern coastal ranges of New South Wales and adjacent areas in eastern Victoria.

NOTES Widely cultivated as a street and ornamental tree for its beautiful upper smooth bark, rich green foliage and profusion of flowers that appear in spherical masses because of the great number of buds in each cluster.

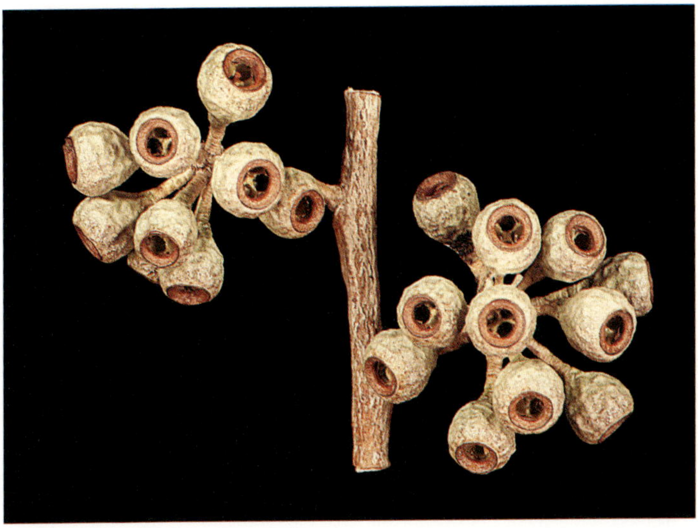

Eucalyptus dives

BROAD-LEAVED PEPPERMINT, PEPPERMINT, BLUE PEPPERMINT

DESCRIPTION Small to medium-sized tree. **Bark** rough, finely fibrous, finely longitudinally fissured, grey or grey-brown. **Juvenile leaves** stalkless, opposite for many pairs, ovate, heart-shaped, to 12 x 7 cm, greenish grey or bluish and often glaucous. **Adult leaves** stalked, lanceolate to broad-lanceolate, to 15 x 3.3 cm, concolorous, slightly glossy, green, with strong peppermint smell when crushed. **Buds** in axils of leaves, in 11s or more, stalked, club-shaped, to 0.6 x 0.4 cm; operculum hemispherical or conical. **Flowers** white, September to October. **Fruit** stalked, cup-shaped, hemispherical or funnel-shaped, to 0.7 x 0.7 cm.

DISTRIBUTION Widespread on the central and southern tablelands of New South Wales and over much of central and eastern Victoria.

NOTES Among the peppermints it is recognised by the large, broad, usually glaucous juvenile leaves usually held in opposite pairs. It has been harvested commercially since the nineteenth century for its essential oil, cineole.

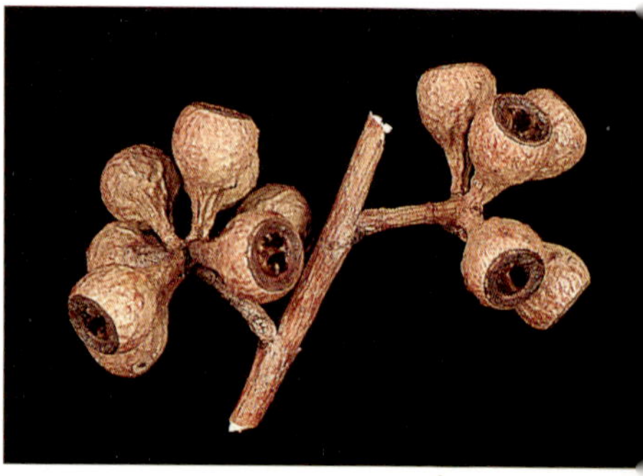

Eucalyptus amygdalina

BLACK PEPPERMINT

DESCRIPTION Small to medium-sized tree. **Bark** rough, fibrous, with fine longitudinal fissures, dark grey-brown. **Juvenile leaves** stalkless, opposite for many pairs, elliptical, to 8 x 2 cm, grey-green. **Adult leaves** stalked, narrow-lanceolate, often with a fine, hooked point on the end, to 11.5 x 0.9 cm, concolorous, green, with strong peppermint smell when crushed. **Buds** in axils of leaves, in 11s or more, stalked, club-shaped, to 0.5 x 0.3 cm; operculum, hemispherical to conical. **Flowers** white, November to January. **Fruit** shortly stalked or stalkless, cup-shaped, hemispherical or funnel-shaped, to 0.7 x 0.7 cm.

DISTRIBUTION Endemic to Tasmania; widespread in the central, northern and eastern regions of the State.

NOTES It is distinguished from the other narrow-leaved peppermint in Tasmania (*E. pulchella*) by the rough bark. Early leaves lack oil glands.

Eucalyptus pulchella

WHITE PEPPERMINT, NARROW-LEAVED PEPPERMINT

DESCRIPTION Small to medium-sized tree. **Bark** smooth, white, pale yellowish or greyish; some large, old trees may have a short, fibrous stocking of rough bark. **Juvenile leaves** stalkless, at first lanceolate to narrow-lanceolate, soon becoming linear, to 7 x 0.4 cm, blue-green to green. **Adult leaves** stalked, linear to narrow-lanceolate, often tapering to a fine point, sometimes hooked, to 10 x 0.7 cm, concolorous, green or grey-green. **Buds** in axils of leaves, in 15s or more, stalked, club-shaped, to 0.5 x 0.3 cm; operculum hemispherical. **Flowers** white, December to March. **Fruit** stalked, crowded, cup-shaped to truncate-globose, to 0.6 x 0.7 cm.

DISTRIBUTION Confined to dolerite-derived soils of the lowlands and hills of mid-eastern and south-eastern Tasmania.

NOTES A popular ornamental with its fine, narrow leaves. It differs from the other narrow-leaved peppermint in Tasmania by the smooth bark.

Eucalyptus risdonii

RISDON PEPPERMINT

DESCRIPTION Shrub, mallee or small tree. **Bark** smooth, or imperfectly shed in larger, old trees, light grey to creamy white; often with characteristic scars encircling the stems formed by the bases of fallen connate juvenile leaves. **Juvenile leaves** at first opposite and stalkless for many pairs, ovate then becoming connate (the leaf pair joined around the stem), the pair to 6 x 5 cm, very glaucous. **Adult leaves** (not often developed), stalked, elliptical-lanceolate, to 10 x 2 cm, blue-grey. **Buds** in axils of leaves, in 7s or more, stalked, club-shaped, glaucous, usually warty, to 0.7 x 0.4 cm; operculum hemispherical, sometimes slightly pointed. **Flowers** white, October to December. **Fruit** shortly stalked, cup-shaped to funnel-shaped or hemispherical, glaucous, to 1 x 0.9 cm.

DISTRIBUTION Of very restricted occurrence in south-eastern Tasmania, known only from the eastern side of the Derwent River, on low mudstone hills from Risdon to Cambridge.

NOTES A popular ornamental because of the attractive round to ovate, glaucous juvenile leaves. These provide a botanical curiosity, as they are in joined pairs.

Eucalyptus coccifera

TASMANIAN SNOW GUM

DESCRIPTION Shrub or small to medium-sized tree. **Bark** smooth throughout, shedding in long strips exposing cream to yellow or pinkish bark which weathers to light and dark grey. **Juvenile leaves** stalkless, opposite for many pairs, broad-elliptical to round, with small points at the tip, to 5 x 4 cm, dull, bluish green; young branchlets warty, usually glaucous, red beneath the white, waxy bloom. **Adult leaves** stalked, elliptical to lanceolate, with hooked tips on ends, to 10 x 2 cm, concolorous, dull or glossy, bluish grey or green. **Buds** in axils of leaves, in 3s to 7s (9s), stalked, club-shaped, glaucous, wrinkled and warty, usually angular, to 1 x 0.6 cm; operculum flattened, much shorter than base, very warty. **Flowers** white, January to February. **Fruit** stalkless or shortly stalked, funnel-shaped or hemispherical, usually glaucous, to 1.1 x 1.3 cm.

DISTRIBUTION Confined to high altitude regions of Tasmania, from Mt Wellington near Hobart extending to the central plateau.

NOTES It differs from the 'true' snow gums by the smaller leaves that lack longitudinal venation, and by the buds being in 3s or 7s, rarely more, with a flattish, warty operculum.

182

Eucalyptus fraxinoides

WHITE ASH, WHITE MOUNTAIN ASH

DESCRIPTION Medium-sized to tall tree. **Bark** rough at base for a few metres, compact, hard, fissured, dark grey to black; shedding above in long strips to leave a smooth, yellowish white surface often with insect 'scribbles'. **Juvenile leaves** stalked, ovate or broad-lanceolate, often curved, to 20 x 6.5 cm, pendulous, dull, blue-green. **Adult leaves** stalked, lanceolate or curved, to 16 x 2.5 cm, concolorous, glossy, green. **Buds** in axils of leaves, in 7s to 11s, stalked, club-shaped, often warty, to 0.7 x 0.3 cm; operculum conical to beaked. **Flowers** white, December to January. **Fruit** stalked, urn-shaped or occasionally truncate-globose, to 1.1 x 1.1 cm.

DISTRIBUTION Southern highlands of New South Wales and the far south-east coastal ranges extending just into the far eastern corner of Victoria in the Howe Ranges.

NOTES It is conspicuous in the high country for the black butt and dense array of 'scribbles' on the smooth bark above. Often in dense, pure stands.

183

Eucalyptus oreades

BLUE MOUNTAINS ASH, WHITE ASH, SMOOTH-BARKED MOUNTAIN ASH

DESCRIPTION Medium-sized to tall tree. **Bark** rough in a basal stocking only, loose or compact, grey; shedding above in long ribbons which remain irregularly attached, to leave a smooth, white, cream or greyish surface. **Juvenile leaves** stalked, ovate or broadly curved and oblique, to 21 x 10 cm, pendulous, dull, blue-green to glaucous. **Adult leaves** stalked, lanceolate or curved, to 17 x 2.3 cm, concolorous, glossy, green. **Buds** in axils of leaves, in 7s, stalked, broadly spindle-shaped and curved, to 0.7 x 0.4 cm; operculum conical (pointed). **Flowers** white, January to February. **Fruit** stalked, cylindrical, barrel-shaped or cup-shaped, to 1 x 1 cm.

DISTRIBUTION Disjunct, widely separated occurrences from the Blue Mountains north to the Macpherson Range in the New South Wales–Queensland border region.

NOTES It differs from the other blackbutt ash, *E. fraxinoides*, in the long slabs and ribbons of partly shed bark at the butt and in the thick-rimmed fruit.

Eucalyptus luehmanniana

YELLOW-TOP MALLEE ASH

DESCRIPTION Mallee. **Bark** smooth, whitish. Young stems and branchlets often glaucous. **Juvenile leaves** stalked, ovate to broad-lanceolate or broadly sickle-shaped and oblique, to 20 x 9 cm, dull, grey to glaucous. **Adult leaves** stalked, lanceolate or sickle-shaped, to 18 x 3.5 cm, thick, glossy, green; young branchlets strongly angled. **Buds** in axils of leaves, in 11s or more, stalked, double-conic or club-shaped, angular, yellow or glaucous, to 1.5 x 0.5 cm; operculum beaked, pointed. **Flowers** white, August to November. **Fruit** stalked, cup-shaped, urn-shaped or almost barrel-shaped, to 1.3 x 1 cm.

DISTRIBUTION Very restricted in coastal scrub around Sydney.

NOTES A striking, coarse-leaved mallee belonging to the blue-leaf ash group. *E. langleyi*, a somewhat similar green-leaf mallee ash from south of Nowra, has glossy green juvenile leaves and less coarse buds and fruit, which place it in the *E. stricta* group.

Eucalyptus pauciflora

SNOW GUM, CABBAGE GUM,
WHITE SALLY

DESCRIPTION Small to medium-sized, rarely tall tree. **Bark** smooth throughout, white or greyish white, often with insect 'scribbles'. **Juvenile leaves** stalked, ovate or broadly curved and oblique, to 18 x 7.5 cm, pendulous, dull, blue-green to glaucous. **Adult leaves** stalked, lanceolate to broad-lanceolate, to 16 x 2.7 cm, thick, concolorous, glossy, green or blue-green, with parallel venation. **Buds** in axils of leaves, in 11s or more, with short, stout stalks, club-shaped, often warty, sometimes glaucous, to 0.9 x 0.5 cm. **Flowers** white, October to January. **Fruit** almost stalkless, cup-shaped to funnel-shaped, to 1.2 x 1.1 cm.

DISTRIBUTION Occurring on mountains and tablelands from far south-eastern Queensland through New South Wales to Victoria and Tasmania; also from lower country near Bega in New South Wales, in Tasmania, south-western Victoria and one population in far south-eastern South Australia.

NOTES The well-known and widely planted snow gum, it varies often from site to site. The more conspicuous forms are the angular-budded subsp. *debeuzevillei*, the very glossy-leaved subsp. *acerina*, the more delicate-budded, small-leaved subsp. *niphophila*, and the sessile-fruited subsp. *hedraia*.

Eucalyptus delegatensis

ALPINE ASH, GUM-TOPPED STRINGYBARK, WHITE-TOP

DESCRIPTION Medium-sized to very tall tree. **Bark** rough on lower part of trunk, thick, fibrous, longitudinally fissured, grey to brown; upper bark smooth, whitish or yellowish grey, often with insect 'scribbles'. **Young branchlets** glaucous, or red where the wax has been abraded. **Juvenile leaves** stalked, ovate to broadly curved and oblique, to 25 x 10 cm, pendulous, dull, blue-green to glaucous. **Adult leaves** stalked, lanceolate or sickle-shaped and oblique, to 18 x 3 cm, concolorous, glossy, green. **Buds** in axils of leaves, in 7s to 15s, stalked, club-shaped, to 0.6 x 0.4 cm; operculum hemispherical with a small point or shallowly conical. **Flowers** white, December to March. **Fruit** stalked, barrel-shaped or rarely hemispherical, to 1.2 x 0.9 cm.

DISTRIBUTION Occurs in sub-alpine altitudes of southern New South Wales and eastern Victoria as far west as Mt Macedon. A related subspecies, *tasmaniensis*, is common over a large area of Tasmania at low and high altitudes.

NOTES Subspecies *tasmaniensis* carries its rough bark on the whole of the trunk and onto the larger limbs. The juvenile leaves are rounder and the stems warty-glandular in contrast to the mainland subspecies.

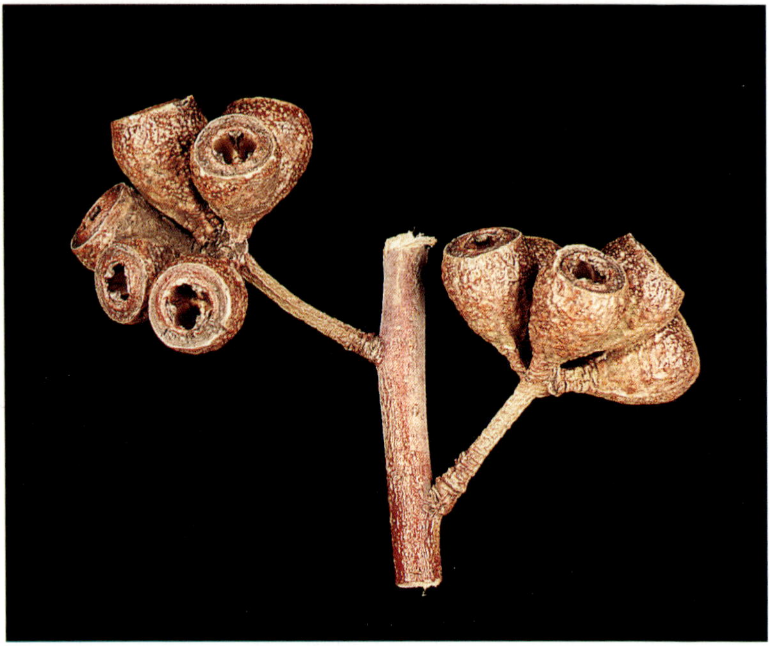

Eucalyptus sieberi

SILVERTOP ASH, COAST ASH, IRONBARK

DESCRIPTION Medium-sized to tall tree. **Bark** of saplings and young trees rough, persistent on most of the trunk, thin, flaky, orange-brown; in the mature tree, persistent on part or whole trunk and larger limbs, compact and coarsely furrowed, dark grey-brown to grey-black; smooth, whitish on smaller limbs. **Juvenile leaves** stalked, ovate or broadly curved and oblique, to 17 x 7.5 cm, pendulous, dull, blue-green to glaucous; young branchlets shiny red overlaid with a whitish wax. **Adult leaves** stalked, lanceolate, to 15 x 2.8 cm, concolorous, glossy, green. **Buds** in axils of leaves, in 7s to 15s, stalked, club-shaped, to 0.7 x 0.35 cm; operculum hemispherical or shallowly conical. **Flowers** white, September to January. **Fruit** stalked, funnel-shaped, to 1.1 x 0.9 cm; valves 3.

DISTRIBUTION Tablelands and coast of the southern half of New South Wales (from north-west of Wyong) to eastern Victoria, with an outlier in the Pyrete Range north-west of Melbourne. Also in north-eastern Tasmania.

NOTES The commonest ash species of south-eastern Australia, it is harvested widely for timber and fibre. It regrows in dense stands from seed on roadsides and exposed forest edges. The fruit usually have 3 valves, in contrast to the following related species.

Eucalyptus consideniana

YERTCHUK

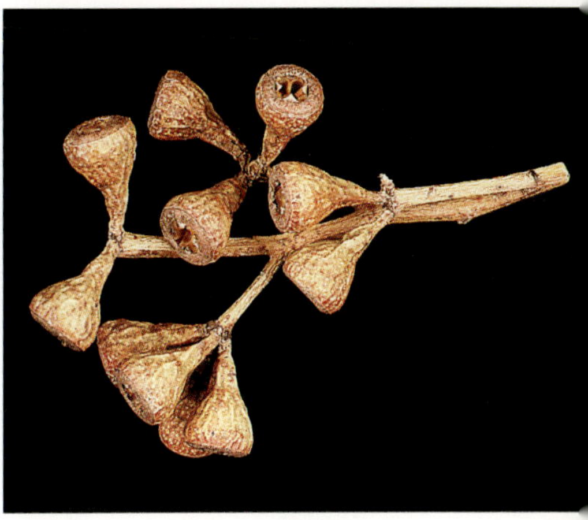

DESCRIPTION Small to medium-sized tree. **Bark** rough, fibrous moderately thick, finely furrowed, grey over yellow-brown. **Juvenile leaves** stalked, ovate or broadly curved and oblique, to 17 x 8 cm, pendulous, dull, blue-green, but not glaucous. **Adult leaves** stalked, lanceolate, to 14 x 2.5 cm, concolorous, glossy, green to grey-green. **Buds** in axils of leaves, in 11s to 15s, stalked, club-shaped, to 0.7 x 0.4 cm; operculum hemispherical with a small point. **Flowers** white, October to December. **Fruit** stalked, funnel-shaped or pear-shaped, to 1 x 0.8 cm; valves 4.

DISTRIBUTION Tablelands and coast of the southern half of New South Wales extending to the Gippsland region of Victoria.

NOTES An ash related to *E. sieberi*, it grows on poorer soils and can be distinguished in the field by the finely fibrous bark, the non-glaucous juvenile growth, and 4-valved fruit.

Eucalyptus stenostoma

JILLAGA ASH

DESCRIPTION Medium-sized tree. **Bark** rough at the base for several metres, compact, hard, finely fissured, dark grey to almost black; smooth, yellowish white above. **Juvenile leaves** stalked, ovate to broad-lanceolate, to 18 x 6 cm, pendulous, dull, blue-green to glaucous. **Adult leaves** stalked, lanceolate or curved, to 19 x 3 cm, concolorous, glossy, green. **Buds** pendulous in axils of leaves, in groups of more than 20, on long, slender stalks, club-shaped, usually glaucous, to 0.6 x 0.3 cm; operculum hemispherical to conical. **Flowers** white, summer. **Fruit** pendulous on long, slender stalks, globose with a small opening, usually glaucous, sometimes with white spots, to 1 x 1 cm.

DISTRIBUTION Of very restricted distribution on tablelands and mountains west of Bodalla in New South Wales.

NOTES A species not closely related to any other. Recognised by the delicate, pendulous, glaucous buds and almost spherical fruits.

Eucalyptus haemastoma

SCRIBBLY GUM

DESCRIPTION Small to medium-sized tree. **Bark** smooth, whitish, grey or yellowish, usually with insect 'scribbles'. **Juvenile leaves** stalked, ovate or broadly curved and oblique, to 22 x 8 cm, pendulous, blue-green. **Adult leaves** stalked, broad- lanceolate or curved, to 15 x 3 cm, concolorous, slightly glossy, green. **Buds** in axils of leaves, in 7s or more, stalked, club-shaped, to 0.8 x 0.4 cm; operculum hemispherical or conical. **Flowers** white, September to December. **Fruit** stalked, funnel-shaped or cup-shaped, to 0.9 x 0.9 cm.

DISTRIBUTION Restricted to the coastal plains and hills around Sydney.

NOTES One of three scribbly gums, it has the largest buds and fruit. It differs from the tableland scribbly gum species, *E. rossii*, which has the narrow juvenile leaves and smaller buds and fruit.

191

Eucalyptus planchoniana

NEEDLEBARK STRINGYBARK, PLANCHON'S STRINGYBARK, BASTARD TALLOWWOOD

DESCRIPTION Small to medium-sized tree. **Bark** rough, thick and spongy, relatively short-fibred, breaking easily into fragments, red-brown to grey-brown. **Juvenile leaves** stalked, ovate or broadly curved and oblique, to 20 x 10 cm, pendulous, discolorous, blue-green. **Adult leaves** stalked, broad-lanceolate or broadly curved and oblique, to 17.5 x 3 cm, concolorous, light green to bluish green. **Buds** in axils of leaves, in 7s, with stout stalks, elongated and cylindroid, strongly ribbed, to 3 x 0.9 cm; operculum conical or slightly beaked. **Flowers** white, November to January. **Fruit** shortly stalked, barrel-shaped to truncate-globose, ribbed, to 2.6 x 2.6 cm.

DISTRIBUTION Of scattered occurrence in south-east Queensland including Plunkett, North Stradbroke and Moreton Islands, south of Brisbane and inland north-west of Gatton; also Gibraltar Range and north coast region of New South Wales; always on infertile sandy soils.

NOTES Easily recognised by the reddish rough bark, bluish curved leaves and the largest fruit of eucalypts in eastern Australia.

Eucalyptus jacksonii

RED TINGLE

DESCRIPTION Tall tree, often prominently buttressed. **Bark** rough, shallowly furrowed, stringy, grey over red-brown. **Juvenile leaves** stalked, ovate, to 17 x 10 cm, slightly glossy, green. **Adult leaves** stalked, broad-lanceolate to ovate, oblique, to 11 x 2.5 cm, discolorous, slightly glossy, dark green, with numerous oil glands. **Buds** in axils of leaves, in 7s, stalked, club-shaped or ovoid to broadly spindle-shaped, to 0.7 x 0.4 cm; operculum obtusely conical. **Flowers** white, January to March. **Fruit** stalked, truncate-globose, to 0.8 x 1 cm.
DISTRIBUTION Of restricted occurrence west, north and east of Walpole in the far south-west of Western Australia.
NOTES One of the tallest eucalypts and often with a massive butt, although very few giants remain. It differs from the unrelated yellow tingle (q.v.) in the wood colour.

Eucalyptus marginata

JARRAH

DESCRIPTION Varying from mallee form to a tall tree. **Bark** rough, fibrous, usually held in flat, longitudinal strips, grey to dark grey. **Juvenile leaves** stalkless, ovate to broad-lanceolate, held horizontally, to 15 x 7 cm, slightly glossy, green to light green. **Adult leaves** stalked, lanceolate to broad-lanceolate or curved, to 13 x 3 cm, discolorous, glossy, dark green. **Buds** in axils of leaves, in 7s or more, stalked, cylindrical to spindle-shaped, to 1.7 x 0.5 cm; operculum conical to horn-shaped. **Flowers** white (rarely pink), September to January. **Fruit** stalked, globose to barrel-shaped, to 1.6 x 1.5 cm.

DISTRIBUTION South-western corner of Western Australia, occurring as a mallee or small tree in its northernmost distribution (Mt Lesueur area) and in southern coastal heaths, and as a tall forest tree in the far south-west; with small outliers at Jilakin Rock and Tutanning.

NOTES The most widespread of the commercially important timber trees of Western Australia. It has multiple uses, from heavy construction to high-quality furniture and railway sleepers. It usually has dark green leaves, but a 'blue-leaved' form (subsp. *thalassica*) has recently been described.

Eucalyptus diversifolia

SOAP MALLEE, COASTAL WHITE MALLEE

DESCRIPTION Small to tall mallee or sometimes a prostrate shrub. **Bark** smooth, grey, whitish, brown or pinkish grey. **Juvenile leaves** stalkless, opposite for many pairs, stem-clasping, ovate or elliptical to broad-lanceolate, to 12 x 8 cm, slightly glossy, green to blue-green. **Adult leaves** stalked, lanceolate, somewhat erect, to 12 x 2.3 cm, concolorous, dull grey-green to blue-green. **Buds** in axils of leaves, in 7s to 11s or more, stalkless or shortly stalked, double-conic, to 1 x 0.6 cm; operculum conical. **Flowers** white, September to January. **Fruit** stalkless or shortly stalked, cup-shaped to hemispherical or obconical, to 1.2 x 1.5 cm; disc broad, level to slightly ascending.

DISTRIBUTION Of coastal and subcoastal distribution from far western Victoria (Cape Nelson), South Australia (to north of Minnipa), Kangaroo Island, and along the shore of the Great Australian Bight into Western Australia; occurring on sand or limestone.

NOTES The extremes in form can be seen on Kangaroo Island, where it varies from a prostrate shrub about 30 cm high on exposed coastal cliffs, to a robust mallee to 15 m high in sheltered situations.

Eucalyptus buprestium

BALL-FRUITED MALLEE

DESCRIPTION Small mallee. **Bark** smooth, greyish. **Crown** usually dense and round. **Juvenile leaves** stalked, ovate, to 9 x 4.5 cm, pendulous, dull, blue-green. **Adult leaves** stalked, narrow-lanceolate, with hooked tips, to 7 x 1 cm, concolorous, slightly glossy, green, with dense venation and virtually no oil glands visible. **Buds** in axils of leaves, in 11s or 13s, stalked, ovoid, to 0.7 x 0.5 cm; operculum hemispherical with a small point. **Flowers** white, November to April. **Fruit** stalkless or shortly stalked, globose, with a relatively small opening, slightly ribbed, to 2.3 x 2.5 cm.

DISTRIBUTION Coastal and sub-coastal sandplains and low hills in the south-west of Western Australia, particularly from the Stirling Range to Bremer Bay.

NOTES An attractive small mallee with dense crown and large globular fruit.

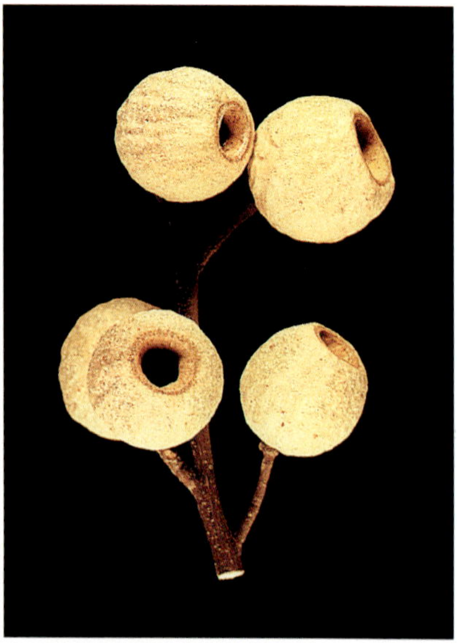

Eucalyptus sepulcralis

WEEPING MALLEE

DESCRIPTION Tall, slender-stemmed mallee. **Bark** smooth, grey and pinkish grey. **Branchlets** and often buds and fruits glaucous. **Juvenile leaves** stalkless and opposite for many pairs gradually becoming stalked and alternate, broad-lanceolate to ovate, pendulous, to 14 x 5 cm. **Adult leaves** stalked, narrow-lanceolate, to 10.5 x 1.2 cm, concolorous, glossy, green to dark green. **Buds** in axils of leaves, in 7s, on long stalks, broadly spindle-shaped or club-shaped, to 1.6 x 0.7 cm; operculum conical to hemispherical with a small point. **Flowers** pale yellow, September to January. **Fruit** stalked, barrel-shaped to urn-shaped, to 3.5 x 2.4 cm.

DISTRIBUTION Known only from near the eastern end of the Fitzgerald River National Park of Western Australia, on lateritic sandplains and low hills.

NOTES Of extraordinary habit, with its tall thin stems and finally pendent crowns. It is conspicuous in its native habitat by emerging well above the heathy understorey.

Eucalyptus aquilina

CAPE LE GRAND MALLEE

DESCRIPTION Mallee. **Bark** smooth, grey, pinkish grey or creamy white. **Juvenile leaves** stalked, broadly elliptical, with crinkly and toothed edges, pendulous, to 16 x 8 cm, dull, blue-green. **Adult leaves** stalked, lanceolate or curved, to 13 x 2.5 cm, concolorous, slightly glossy, green to dark green. **Buds** in axils of leaves, in 3s, on broad, flattened straps, stalkless, double-conic, to 3.5 x 3.5 cm; operculum shallowly conical. **Flowers** white, April to June. **Fruit** stalkless, broadly funnel-shaped, smooth or faintly ribbed, to 2.5 x 5 cm; disc broad; with large prominent lobes covering the valves.

DISTRIBUTION Very restricted in the south coast region of south-central Western Australia, in the shallow valleys, creek beds and hillsides of the Mount Le Grand, Frenchman's Peak and Thistle Cove areas, and also on Sandy Hook Island.

NOTES Originally collected by Robert Brown on the Flinders voyage of 1802, it was not collected again until the early 1970s, when it was officially named. Notable for the beautifully sculptured fruit.

198

Eucalyptus coronata

CROWNED MALLEE

DESCRIPTION Small mallee. **Bark** smooth, pinkish grey, grey and light grey over orange. **Juvenile leaves** stalkless at first, soon becoming stalked, broadly elliptical, to 15 x 11 cm, light green. **Adult leaves** stalked, lanceolate, to 12 x 2.7 cm, concolorous, slightly glossy, green to blue-green. **Buds** in axils of leaves, in 3s, on broad, flattened straps, stalkless, double conic, prominently ribbed, to 3 x 2.6 cm; operculum conical. **Flowers** white, July to August. **Fruit** stalkless, broadly funnel-shaped, ribbed, to 2.3 x 5.2 cm; disc broad; with prominent lobes covering the valves.

DISTRIBUTION Known only on hills from Middle Mt Barren to East Mt Barren, west of Hopetoun on the southern coast of Western Australia.

NOTES A very rare species notable for its coarsely ribbed, sculptured fruit.

199

Eucalyptus preissiana

BELL-FRUITED MALLEE

DESCRIPTION Low, sprawling mallee. **Bark** smooth, grey over pinkish grey or cream. **Juvenile leaves** opposite and stalkless for many pairs, elliptical to ovate, to 15 x 8 cm, with toothed edges, dull, light green or blue green; stems hairy. **Leaves** on mature plant in late juvenile phase, becoming alternate and shortly stalked, elliptical to broad-lanceolate, to 11 x 5.5 cm, thick, concolorous, dull, light green. **Buds** in axils of leaves, in 3s, on broad, flattened straps, stalkless or with short, stout stalks, ovoid, to 2.6 x 1.9 cm; operculum conical. **Flowers** pale yellow, August to November. **Fruit** stalkless or with short, stout stalks, funnel-shaped to cup-shaped or bell-shaped, to 2.8 x 3.2 cm; disc broad, descending or level, slightly lobed.

DISTRIBUTION Southern Western Australia, widespread in coastal and subcoastal sandplains including the Stirling Range and Fitzgerald River National Parks, east to Stokes Inlet.

NOTES A favourite ornamental because of its low sprawling form and spectacular yellow flowers. The eastern form of this species, from Starvation Boat Harbour almost to Esperance, has recently been published as subsp. *lobata*.

200

Eucalyptus insularis

North Twin Peak Island Mallee

Description Small, slender-stemmed to moderately tall mallee. **Bark** of tall mallees, fibrous, red-brown at butt, otherwise smooth, red-brown, light grey, yellow-green, or greenish grey. **Juvenile leaves** at first stalkless, opposite, elliptical, with toothed edges, to 7 x 3 cm, soon becoming stalked and alternating, lanceolate, to 7 x 1.5 cm, dull, green; stems warty. **Adult leaves** stalked, lanceolate, narrow-lanceolate or curved, to 7.5 x 0.8 cm, concolorous, slightly glossy, light green. **Buds** in axils of leaves, in 7s or more, stalked, pendulous, broadly club-shaped, to 0.7 x 0.4 cm; operculum hemispherical with a small central knob. **Flowers** white; flowering period not known. **Fruit** stalked, pendulous, barrel-shaped, to 0.8 x 0.6 cm.

Distribution Of very restricted occurrence in the south-central coastal region of Western Australia, known only from North Twin Peak Island, Mount Le Grand and a nearby granite rock.

Notes A tall robust mallee in its island occurrence and a delicate, dense-crowned shrub on the mainland. It resembles no other eucalypt.

GLOSSARY

adult leaves the final growth phase of the leaves in most eucalypts, although some species are reproductively mature in more juvenile phases

alternate the leaves alternate along the branchlet, not opposite

axil angle between the branchlet and the leaf stalk

axillary position of the buds in the angle between the leaf stalk and the stem

bark

bloodwood – rough bark held in distinct small flakes or tesserae

box – rough bark, often abrading in small flakes

fibrous – rough bark held in fibres

ironbark – rough, thick, hard, furrowed bark

minniritchi – rough, partly shed reddish curled outer bark, inner greenish

powdery – smooth bark that is powdery to the touch

ribbony – partly shed ribbons of bark held in crown

scribbly – irregular insect larvae tracks on smooth bark

smooth – outer bark shed annually to leave smooth surface

stringy – rough bark held in long thick 'strings'

tessellated – rough bark held in distinct, firm flakes (particularly most bloodwoods and some ghost gums, and to some extent box species)

bloodwoods a large group of species usually recognised by the tessellated rough bark and terminal bud clusters

compound inflorescence the bud clusters occur in groups, not singly

concolorous the leaf has the same colour both sides

connate opposite leaf pairs that are joined at their bases around the branchlet

disc part of the rim of the fruit from the staminal ring to the ovary roof

discolorous the leaf is darker on the upper side

disjunct occurring in widely separated areas, not of continuous distribution

elliptical shaped like an ellipse but often pointed at one or both ends

endemic when the natural distribution of a species is restricted to a certain defined area

flower of the eucalypt strictly the stamens provide the colour

ghost gums a small group of species with notably smooth white bark although rough at the base in some species

gimlet a tree form only in Western Australia notable for the trunk which has spiral flutings

glaucous covered with a white wax on the surface or bluish grey in colour

globose more or less spherical

inflorescence bud clusters which may occur singly or in pairs in the axils, be on extended stalks in the axils or may be several to many, clustered at the leafless ends of the branchlets

intermediate leaves the growth phase between the juvenile and the adult leaves

juvenile leaves the growth phase of

the leaves between the seedling and the intermediate leaves

lanceolate shaped like a lance head, wider at the base and tapering to the tip

linear very narrow, usually with +/- parallel sides

mallee the growth form of many small eucalypts – i.e. a multi-stemmed shrub

mallet small to medium-sized tree often in pure stands with conspicuous terminal crown and steep branching habit – a Western Australian term

monocalypt a large subgroup of the eucalypts (pp. 185–221) which have a single bud cap such that the side of the bud is unscarred; mostly coastal and subcoastal species

oil glands seen in the leaves when held to the light

operculum the flower bud cap

opposite leaves are paired and opposite on the branchlet

ovate egg-shaped in outline with broadest part towards the base

ovoid egg-shaped

peduncle stalk of an individual bud cluster

peltate when the leaf stalk attaches to the blade on the underside near the edge

pith glands occur in the medulla or pith of the branchlets and seen by splitting a branchlet longitudinally, often more prominent at the nodes

scabrid rough to the touch

seedling the growth phase of the leaves between the cotyledons and the juvenile leaves

stamens the male reproductive part of a flower, a filament surmounted by an anther

style the needle-like upper extension of the ovary down through which the pollen tube passes in the fertilisation process. In the dehiscence of the fruit it splits longitudinally and the fragments may persist and extend above the rim after seed is shed.

terminal the inflorescences are at the more or less leafless ends of the branchlets, not in the axils

truncate-globose more or less spherical but cut off at the top

truncate-ovoid broadly egg-shaped but cut off at the top

valve the tips of the ovary roof in the eucalypt fruit that extend upwards, particularly conspicuous when they exceed the rim

wing-seeded the seed has a transparent wing extending above the body of the seed – a character of most bloodwoods

REFERENCES

Boland, D.J., Brooker, M.I.H., Chippendale, G.M., Hyland, B.P.M., Johnston, R.D., Hall, N., Kleinig, D.A. and Turner, J.D. (1984). *Forest Trees of Australia*, Thomas Nelson and CSIRO, Melbourne.

Brock, J. (1992). *Native Plants of Northern Australia*, Reed Books, Sydney.

Brooker, M.I.H. and Kleinig, D.A. (1983). *Field Guide to Eucalypts*, Vol. 1, Inkata Press, Melbourne.

Brooker, M.I.H. and Kleinig, D.A. (1990). *Field Guide to Eucalypts*, Vol. 2, Inkata Press, Melbourne.

Brooker, M.I.H. and Kleinig, D.A. (1994). *Field Guide to Eucalypts*, Vol. 3, Inkata Press, Melbourne.

Brooker, M.I.H. and Slee, A.V. (1996). *Eucalyptus*, in *Flora of Victoria* Government Printer, Melbourne.

Chippendale, G.M. (1988). *Flora of Australia*, Vol. 19, Australian Government Publishing Service, Canberra.

Cronin, Leonard (1987). *Key Guide to Australian Wildflowers*, Reed Books, Sydney.

Elliot, W.R. and Jones, D.L. (1986). *Encyclopaedia of Australian Plants*, Vol. 4, Lothian, Melbourne.

Hill, K.D. (1991). *Eucalyptus*, in *Flora of New South Wales*, Vol. 2, New South Wales University Press, Sydney.

Stanley, T.D. and Ross, E.M. (1986). *Eucalyptus*, in *Flora of South-eastern Queensland*, Vol. 2, Queensland Department of Primary Industry, Brisbane.

Wheeler, J.R. (ed.) (1992). *Flora of the Kimberley Region*, Department of Conservation and Land Management, Perth.

Index

Botanical names

Common names